HINDRANCES FOR THE PROMOTION OF ECOLOGY: A CROSS CULTURAL STUDY

A THESIS SUBMITTED FOR THE AWARD OF THE DEGREE OF
DOCTOR OF PHILOSOPHY

BY

MUKTIKANTA ACHARYA, M.A

UNDER THE GUIDANCE OF

DR. RASIKANANDA MOHANTY

SUBMITTED TO
FAKIR MOHAN UNIVERSITY, VYASA VIHAR, BALASORE
ODISHA
2014

CONTENTS

Chapters	Name	Pages
	Preface	
Chapter-I	Outlines of Environmental awareness and Eco-System	1
Chapter-II	Metaphysical models and Conflicting cultural patterns: It's impact on Environment and Eco-System	45
Chapter-III	Philosophical thinking of modern man in terms of mechanistic view	62
Chapter-IV	Philosophical thinking of man in terms of Environment and Eco-System	122
Chapter-V	Comparative study of Chapter-III and IV	154
Chapter-VI	Dangerous effects due to cross thinking	172
		182

CHAPTER-I

OUTLINES OF ENVIRONMENTAL AWARENESS & ECO-SYSTEM

OUTLINES OF ENVIRONMENTAL AWARENESS & ECO-SYSTEM

The present day world is facing innumerable environmental as well as ecological problems due to rapid growth of modern civilization & industrial set up. The man at present is advancing at each moment towards a great destruction for environmental & ecological imbalance. I have constantly tried in my thesis to search a way out of this danger of pollution, by giving a philosophical approach to solve problems of pollution in environment & Eco-system.

ENVIRONMENT AND ECOLOGY:

Before illustrating environmental awareness, we should first of all explain, what is environment. "Environment" is that which affects the life & development of the organism in its natural habitat. It is the totality of everything that is around us; like biological, social, economical, physical or chemical. All the species influence and are influenced coming under it. Environmental concern is a global momentum in the present context. After completion of world war-II, industrial revolution took place in the European countries for healing of economic wounds caused by devastating war without caring for environmental degradation. So many scientific & technological developments took place to increase production by using natural resources, bypassing the environmental consequences. But when different critical situations arose, world thinkers expressed their concern; UNO arranged various conferences to create consciousness for the protection of the environment. All the environmental situations are arising due to excessive use of nature and natural resources. As a result; climatic change, pollutions, loss of wild habitat, health hazards are growing day by day. Therefore, earth summit in 1992 and Johannesburg convention in 2002 emphasized on sustainable development and finding out solution for controlling environmental degradation.

But the term 'Ecology' is derived from the Greek word 'Oikos' which means house. So ecology is the study of the households of the world, both living and non-living. It includes soil, air, water, different micro-organisms, plants, animals and human beings. As the bodies of the living beings are made up of non-living components, they are interdependent on each other. Ecology is the study of inter-relationship among them as well as with environment. It is an interesting subject, even a child will be eager to know about his surroundings; like flowers, plants, insects, mountains and rivers which exist around him. It is also the biology of organisms occurring in natural habitat like land, ocean, pond etc. E.P.Odum, an American ecologist has defined ecology as "the study of the structure and function of nature which includes the living world".[1] It is a kind of science which deals with the common characteristics of living beings; like survival, adaptability and reproduction in the world.

We say non-living elements are subtle elements or Mahabhutas, such as Kshiti (soil or lithosphere), ap (water or hydrosphere) and marut (air or atmosphere) and different changes like diurnal, nocturnal, seasonal and annual are produced out of these elements. In ecology, we study populations, various communities, organisms and biosphere etc. A group of individuals of any kind is called population. And community means population of a particular area which is called habitat. In eco-system, community interacts with abiotic environment.

The term Eco-system was first coined by British ecologist A. G. Transley in 1935. In eco-system atmosphere, lithosphere and hydrosphere interact with each other to make life possible on earth. They can be classified also.

---

1. Kormondy Edward J, Concepts Of Ecology', 4th Edition, PHI Learning Pvt. Ltd. Page-4

CLASSIFICATION OF ATMOSPHERE:

It has four layers. Such as (a) Troposphere (b) Stratosphere (c) Ionosphere (d) Exosphere.

(a) Troposphere

It varies in height according to latitude. It is 16 to 18 km high from the ground at equator, 10 to 12 km at moderate latitude and 8 to 10 km in polar region. Water vapour varies according to tropospheric variation, i.e in lower height water vapour is more and in upper height it is less. Troposphere becomes hot according to radiation of the solar energy and differs in degree in accordance with height. So also the density of water vapour varies, depending on the height.

(b) Stratosphere

It is about 50 to 55 km high, free from cloud and aeroplane fly in this layer. Ozone layer is found in this layer, absorbs the ultraviolet rays of the sun and acts as an umbrella for the earth. A big hole is found in this Ozone layer over Antarctica region due to excessive emission of harmful gasses, which is a serious threat for all the life forms on earth.

(c) Ionosphere

It is started after stratosphere and contains different ionized air. As it reflects short radio waves, long distance communication is possible here.

(d) Exosphere

It is the last part of the atmosphere, so the density of the air is very low. Outer space is started just after it.

Lithosphere:

It is the body of the earth. Scientists say that earth was formed 500-600 crores years ago. It has also three layers. (i) The earth's crust (ii) The mantle (iii) the core

(i) The earth's crust

It is solid in form and it ranges from 16 to 50 km in thickness. Depending upon the nature of the soil, our surface soil stretches from few inches to few feet. Different types of organic activities are done here.

(ii) The mantle

In this part, different metals containing hard rock are found. It stretches up to 2880 km of earth's thickness and the weight is 67% of the earth.

(iii) The Core

It is composed of high density of solid material like nickel and iron and the temperature is about 8000^0c.

Soil is formed due to biological, physical and chemical works on rock. The micro organisms and some vertebrates decompose the organic materials and help in formation of soil. There are different types of soil; like sandy, silty, clayey and loamy etc, which help for the distribution of flora and fauna.

HYDROSPHERE:

Different forms of water come under this sphere. Water is found in three different forms. Like solid (ice), liquid (water) and gaseous (water vapour). Ice is found in polar region, water in ocean, river, lake, stream, pond etc and gaseous form in atmosphere as moisture.

ECOLOGY AND POLLUTION:

Ecology started after appearance of human being on earth. His survival, growth and dependence come under the purview of ecology. Human beings for their food, fodder, cloth, medicine, fuel etc depend on nature. Therefore, they interact with homogeneous species along with other species. So, ecological discipline should be maintained to save billions of organisms in the state of dynamic co-existence. But rampant use of nature by human species is creating problems for other species by disturbing equilibrium on earth.

Different questions come to our mind during the course of discussion on ecology; like natural resources, agricultural systems, fuel production, protection of land erosion, food, fodder, circulation of nutrients between organisms and environment etc. Human beings lived with raw meat, wild fruits and roots of forest in the primitive period. At about twelve thousand years ago, they learnt the use of fire and agriculture for which forests were cleaned resulting influence on environment. Not only for agricultural purpose, forests were continuously cleaned for several purposes; like establishment of villages, towns, cities, transportation facilities, production of energy etc. It was doubled with the increase of human population to meet their growing needs. When men became civilized, they wanted to live comfortable living. So, scientific developments were made to produce more in every sector. Industries were established to meet the growing needs of men. Again to run industries, different metals, fossil fuel, coal were explored. During the work of exploration, many forests were destroyed, wild lives were threatened or destroyed, and metals were depleted by creating a serious threat for the promotion of ecology.

Today, different things are produced for human purpose, which bring harms to the environment and eco-system. Due to excess human activities, the concentration of carbon dioxide is growing in both hydrosphere and atmosphere, which is killing helpful pollinating insects, fishes and other aquatic animals entering into food chain.

So, ecologists are always trying to make a balance between it. Several mathematical theories of functional systems are followed to explain the change and with the change how we can maintain stability. As there is more stress on nature, we give emphasis on individual organisms, population and communities. which play important role in the study of ecology and the scope is broadening with the rise of environmental problems. So it has become interdisciplinary having three levels, (a) Theoretical; perfect knowledge about nature (b) practical; which helps to analyse and understand the problems (c) Sociocultural; people can solve the ecological problems like Chipco Movement in U.P and Gandhamardan Movement in Odisha etc. There are two types of ecology (i) Autecology; ecology of individual species (ii) Synecolgy; ecology of community. As ecology includes both biotic and abiotic substances. So it has four branches, such as;

(1) Abiotic Substances

We all know that biotic bodies are made up of air, water, and soil which are abiotic, take part in metabolism and lastly return to the environment. It has three parts; (a) Climate (b) Water (c) Proteins, Carbohydrates and other humic substances, with which living bodies are formed. Among these, climate has important role in variation of species.

(2) Producers

Basic producers are trees and plant species. They are self nourishing and transform the solar energy into chemical energy with the help of water, carbon dioxide and organic substances like enzymes.

(3) Consumers

There are two types of consumers; herbivores and carnivores. Herbivores are plant eating animals; like cattle, deer, elephants etc; and carnivores are meat eating animals; like lions, tigers etc. Herbivores come under first group and carnivores come under second group.

(4) Decomposers

Their duty is to break the complex organisms of flora and fauna into simple substances. Protozoa and earthworms come under this group. They are the decomposers of second category. Micro organisms are treated as decomposers of first category.

In every eco-system, these four categories are found.

ENERGY AND ECO-SYSTEM:

It is obvious that organic and inorganic substances are complementary to each other. Organic bodies, for their very existence take nutrients from inorganic substances. There are two types of nutrients; macro nutrient and micro nutrient. Macro nutrients are phosphorus, calcium, carbon, hydrogen etc and micronutrients are iron, manganese, magnesium, zinc, cobalt etc. The organic bodies, after death decay in the soil by giving rise to growth of humus, which the plants take for their growth. Fossil fuels, which are used in industries and production of energy mostly, are the organic matter of plant origin. In Eco-system, energy plays an important role. According to E.P. Odum, an American ecologist "the rank of energy is first in classification of Eco-system".[2] The density, diversity, development and function of organisms are determined by it. There are two types of energy; radiant and fixed. Radiant energy is in the form of electromagnetic wave, such as light. Fixed energy is potential chemical energy, present in organic substances, which can be broken down to release the energy content. In eco-system solar energy is the main source of energy. Plants transform the solar energy into chemical energy for photosynthesis. Some chemical energy is used by plants and some are transferred to the eco-system for consumers. Energy always flows from one tropic level to another; it is called thermodynamics. Energy can only be converted from one form to another form, but it cannot be destroyed. While transforming from one level to another, some are degraded into heat and dissipated for their metabolism and some go to next stage of consumers. Some energy also goes to decomposers through dead and living organic matter. It can be represented in this way:-

Solar energy ⟶ Producer ⟶ Primary consumers (Herbivores) ⟶ Secondary Consumers ⟶ (Carnivores) Higher order Consumers (Carnivores)

The energy flow is one directional, which means energy taken by autotrophs does not go back to solar inputs, so also energy taken by herbivores does not go back to autotrophs. There is also another type of energy flow which is two directional or 'Y' shaped, where one arm stands for herbivore food chain and another arm for decomposer food chain. This system was discovered by H.T. Odum in 1956. In this system, two food chains are not isolated from one another. For example, the dead bodies of some herbivore animals become incorporated in the detritious food chain. But they vary in different Eco-system.

---

2. Dash M.C, 'Fundamentals of Ecology', 2nd Edition, 2004, Tata Mc Grow-Hill Companies, Page-22

In production of energy, plants use solar energy and transform it into chemical energy. This is called gross primary production. Some amount of energy is stored in the plant after maintenance, which is called net primary production and appears as new plant biomass. It can be exemplified like this, after yielding in a paddy field, we see grain, stalks, roots, straw, which are net primary production. Paddy plants use energy during their respiration and growth. Dry matter production is equal to photosynthesis with nutrients. So energy, fixed in primary production, which is produced by plants as food material become life support at other tropical level. It plays an important role in Eco-system.

In energy flow system, secondary producers like herbivores, carnivores and decomposers have great importance. Homoeothermic animals, such as, birds and mammals including men and poikilothermic animals, such as fish, amphibians and reptiles come under secondary production system. Herbivores eat plants, some of it are assimilated and utilized in metabolism, growth and reproduction and some are egested. It varies according to aging system. In secondary production, the decomposer organisms differ from herbivores and carnivores in detritious food chain and bring high level of productivity in decomposer tropic system.

Two types of animals are seen in secondary production system, viz; vertebrates and invertebrates. In vertebrate section, cattle, birds, rodents, snakes, frogs etc are included; but in invertebrate section grasshoppers, locusts, beetles (above ground) and earthworms, ants (underground) etc. Due to over grazing, our primary productivity is controlled. Rodents and cattle have deteriorated our primary productivity by over grazing. The invertebrate secondary producers depend upon abundant diversity of plant species and primary productivity. So food chain relationship is very much complex in nature. Sometimes it is seen that, different types of animal depend on single food source; such as rabbit, grasshopper, and cattle take grass as their food. Carnivores take them according to their food habit. It is called food web. Green plants are called producers, they remain in the first tropic level, i.e; insects, rabbits, rodents, deer, cattle are herbivores or primary consumers. Those who eat them like frog, fish eating zooplankton etc. Then another type of consumer who are carnivores, eat the flesh of herbivores and secondary consumers. It can be arranged like this:

Plants ⟶ Herbivores ⟶ Carnivores$_1$ ⟶ Carnivores$_2$

For sustenance of Eco-system, balance should be maintained. A famous ecologist Charles Elton said that "One hill cannot sustain two tigers."[3] A hill has limited primary production. So it cannot sustain so many primary or secondary consumers for the food of tigers. From this it is obvious that, in a small eco-system, many territory or secondary consumers cannot survive.

DIFFERENT ECO-SYSTEMS OF THE WORLD:

Eco-system varies according to climate. We observe two types of eco-system in the world, i.e; terrestrial and aquatic. Terrestrial Eco-system can be classified as forest, grass lands, savanna, tundra, and desert. The region covering a large area is called biome or biochore. The divisional lines of these biomes are parallel to latitude.

CLASSIFICATION OF FORESTS:

There are different types of forests, which are discussed here under.

Tropical rain Forest

Such type of rain forests are seen where temperature and humidity are high. It gets rain fall of 200 to 225 cm over the year. Different types of flora and fauna are found there.

Such types of forests are seen in the Amazon River basin of South America, West Indies, South East Asia, in different parts of Africa and North West Australia. We can see 200-300 species of trees in one square kilometer. Due to heavy rain fall, trees are ever green and tall up to 25-40 meters

Different plant species grow under the shade of these big trees. Soil is very much fertile due to high rate of decomposition. Such types of forests are seen in Gandhamardan hills of Odisha, Kerala and Assam.

---

3. Ibid- Page-83

Tropical rain forests are divided into three classes. (i) Moist tropical forest (ii) Montane sub tropical forest (iii) Montane wet tropical forest.

(i) **Moist tropical forest-** Such types of forests are seen in Assam, West Bengal, Odisha, Kerala and Andaman Island.

(ii) **Montane sub tropical forest-** It is found in northern wet hill forests of West Bengal, the southern sub tropical broad-leaves-hill forests of odisha and Kerala and sub-tropical pine forests of U.P., Himachal Pradesh, Manipur, Assam etc. The leaves of these forests are big in size.

(iii) **Montane wet tropical forests-** It is found in West Bengal, KodaiKanal in Tamilnadu and Udgamandalam in Kerala etc.

Temperate Rain forests

In this forest, climate is very cold. Rain falls in winter and stormy wind blows in this zone. The productivity of this area is very low due to low temperature and lack of nutrient in the soil. The rate of decomposition is slow due to cold climate. These types of forests are seen in North America, South Eastern Australia, West Coast of New Zeland and Tasmania and Southern Chilie.

Tropical and Sub-Tropical Deciduous Forests

In this region rain fall is less, so forest species are low. Summer is too hot and winter is too cold. Trees shed their leaves. Rain falls about 75 to 100 cm per annum. Timbers are available here. The rate of decomposition is high, so also primary productivity.

Temperate Deciduous Forests

Millions of years ago, such types of forests are seen in northern hemisphere. It has three parts; Eastern northern America, Western Europe and Eastern Asia. A moss and a lichen layer grow on rocks, timbers and tree trunks. A herbaceous layer and a shrub layer of about 3 mt high are found. Some trees grow up to 55 mt high. As the climate is cold, the decomposition is slow and land is not fertile.

Montane Coniferous Forests

It is called coniferous, because the leaves are cone bearing in size, i.e; Pines, Firs, Spruce, Hemlock etc. In cold regions these are found. In India these are found in Jammu & Kashmir, Himachal Pradesh, U.P, Assam, and Arunachal Pradesh.

Boreal Coniferous Forests

This type of forest region extends from moderate to extreme cold zone. Precipitation occurs in summer. Red Pine, white Pine, Balsam Fir, Spruce are found in this forest. It has also four layers. Primary productivity is very low due to lack of nutrient. Rodents, Black bears, Moose like animals are seen here. Taiga is the example of this type of forest.

The Chapral Eco-system:

Large shrubs containing sufficient evergreen leaves having waxy materials on its surface are found in this eco-system. Rain falls in winter and summer is dry. Herbs contain thick underground stem in order to tolerate the dry summer. Rodent and reptiles are chief consumers of that area. It is found in Chilie, USA, California, Western Australia and nearing area of Mediterranean Sea.

Tropical savanna:

In tropical rain fed region, such type of areas are seen. It is a type of grass land, which is created either due to climatic condition or destruction of tropical forests. In those areas, various types of climate such as, warm and rainy, cool and dry, hot and dry are seen. Such type of grass lands are seen in India, South America and Africa. The primary productivity is more and decomposition rate is high also. Tall grasses grow in those areas. Varied types of consumers are seen in those areas, like Zebras, elephants, rhinoceroses, lions, leopards and antelopes in African savanna. In Indian savanna, Sal trees and grasses are seen and consumer animals are cattle, jackals, hyenas, rodents. There are five kinds of savannas in India.

(i) High Savanna

In Bramhaputra valley, such type of savanna is found. Trees and grasses grow up to 2-3 mt high.

(ii) Moist Sal Savanna

It is found in Gangetic plain. Sal trees and tall grasses are seen here.

(iii) Low Alluvial Savanna Woodland

Its name is made according to the nature of the soil. The soil is sandy and alluvial. It is found in river flats and Gangetic plain.

(iv) Dry Savanna

Here grasses are available plently. Only a few trees are seen here and there. It is found in Punjab, Hariyana, Bihar, Odisha and eastern part of Tamilnadu.

(v) Saline Alkaline Scrub Savanna

Such type of grass lands are seen in Indo-Gangetic plane.

GRASS LANDS:

We find various types of grass lands in different parts of the world. The grasses grow depending on the precipitation of that area. Where precipitation is more, grasses grow up to 2 mt height. Primary production of that area is 3 tons per acre per year and primary consumers are hoofed animals. Names of the grass lands vary according to region. It is called Prarie in North America, Pampas in Argentina and Steppes in Eurasian region, Pustaza in Hungary and Veldt in South Africa. There are three types of grass lands.

(i) Tropical Grassland

This type of grass land is found 20^0 away from the equator. Rain fall is very low, i.e; 40 to 100 cm per annum. The height of grasses becomes 1.5 to 3.5 mt and consumer animals are deer, antelopes, giraffes and lions.

(ii) Temperate Grasslands

Here rain fall is 25 to 75 cm per annum. It is found in the middle of the continent Europe.

(iii) Alpine Grasslands

It is meadow type and many flowering herbs grow here.

DESERTS:

Deserts are created, where rain fall is less than 25 cm per annum. Two types of deserts are seen in the world. In one type of desert, only thorny bushes are found here and there and another one is fully barren. Some deserts are hot and some are temperate or cold. Sahara in northern Africa and Thar in India are hot deserts. The deserts of Iran and Turkey, some deserts of Argentina and Gobi desert of Mangolia are temperate or cold. The sky is clear, nights are too cold and days are too hot. The soil is sandy; agriculture is possible where water is available. Date palm, Barley, Cotton, Millet are produced in those areas. Primary productivity is very low as water is not available there and consumer animals are Camel and desert rats.

TUNDRA:

In tundra region, the maximum temperature is 10^0c, because it is situated 60^0 north from the equator. Primary productivity is very low as soil is permafrost. Precipitation is very low. Lichens and small flowering herbs are found here and there. The consumer animals are caribou, musk ox, arctic hare and arctic fox etc, but in summer; black flies, mosquitoes and migratory birds are seen there. There are two types of tundra. One is Arctic tundra and other is Alpine tundra. Arctic tundra is real but alpine tundra is tundra like area. The soil of Alpine tundra is better than Arctic tundra.

MANGROVES:

Such type of Eco-system is found in the coastal belts of the world. We find four types of mangroves in the tropical regions, but twenty one types of mangroves are seen in eastern Africa. Their growth becomes 8 to 20 mt and look like bush land of high forest. We find mangroves in Sunderban of West Bengal, Bhitarkanika of Odisha, the Andaman and west coast also. This area is very much productive. Different types of animals and plants are seen there. Sea turtle, Alive Ridley come to Bhitarkanika every year from December to January and March to April to lay eggs. Different types of terrestrial consumers like tigers, hyenas, jackals, deer, turtles, snakes are seen there.

AQUATIC ECO-SYSTEM:

Aqua or water has made this planet living, otherwise it would be like other planets of solar system. There are three types of aquatic Eco-system. (a) Inland water; i,e, rivers, ponds, lakes, springs, etc, (b) Ocean water, which is purely salty. (c) Estuarine water; which is more salty than fresh water. Out of total water found on earth, 97.2% are in ocean and 2.8 are in inland water including snow. The terrestrial Eco-system and aquatic Eco-system are interdependent for their existence, growth and development.

(a) Inland water

There are two types of inland water. (i) Lentic habitat or static water (ii) Lotic habitat or running water.

(i) Lentic water- It is confined in ponds, tanks or ditches. There are three zones in this type of water. First is littoral zone, here light can penetrate up to bottom, because it is a shallow water zone. Second is limnetic Zone, where light can be penetrated, as it is an open water zone. Third is profundal zone, here light cannot penetrate up to the bottom as it is deep water zone.

We see herbs or phytoplankton in littoral zone. Aquatic flower plants like lily, lotus is found here. Snails, dragon flies, hydra, flat worms etc are the consumers of this area.

In limnetic zone primary producers are floating hydrophytes, submerged hydrophytes, and phytoplanktons etc. They vary according to the presence of nutrients, dissolved cases etc.

But in profundal zone, as light cannot pass, producer organisms are not found there. The consumer animals depend on detritus or consumers of litoral or lentic zone.

(ii) Lotic water or Running water- It is found in rivers, streams etc. The force of water depends on its bed, where it is flowing. The flow of the hill side is different from the flow of the plain land. Rivers become wide when many rivulets mix with it. During rainy season, river water become muddy and light cannot penetrate into it. When rivers meet the sea, the land becomes swampy and that is called delta. Here various types of plants and animals, such as, mangroves, crocodiles, fish, phytoplankton, Zooplankton are found. We can see two types of streams, rapid moving and slow moving. In slow moving stream, different kinds of planktons are grown, due to which, different consumers like fish, turtle, water snakes etc are seen. But in rapid moving stream, animals having hooks, which can attach themselves to rocks are seen. In India, different rivers such as Ganga, Yamuna, Bramhaputra, Mahanadi, Godavari, Kaveri ets are flown, which are the life line of our country. They provide water for various purposes; like cultivation, drinking, sanitation, industry etc.

(b) The Ocean Water or Marine Eco-system

We all know that, two third of earth is covered with ocean and one third is land. So the volume of ocean is 15 times more than land. The ocean water is salty due to presence of different salts like sodium, potassium, magnesium, calcium and sulphur etc. which are washed from earth's crust. It is tested that salinity is more at 30^0N and 30^0S latitude. The salinity and temperature vary according to inflow of fresh water. Three types of environments are seen in ocean, the open sea, the ocean depth and the coastal water. Open sea is stretched up to 70 mt depth. Ocean depths are cold, dark and producer animals are not found there. Only carnivores and detritus animals are seen there. The bodies of those animals are flat, eyes are sensitive and they can move to one side. Molluscs and different types of fishes are found. The coastal water extends from sea coast to about 160 km. Here sunlight can penetrate up to 50 mt and primary production is good.

CLIMATIC ZONES:

It is decided according to the rain fall and temperature. We find six climatic zones in our world.

(a) Equatorial Zone- It extends up to 10^0S and 10^0N from the equator. This zone is very hot and full with humidity. So rain fall is very high and average temperature varies between 25^0 and 27^0c.

(b) Tropical zone- It extends from 10^0 to 25^0 on both sides of the equator. The rain fall is high and occurs in hot months. The daily temperature is more than equatorial zone.

(c) The Sub-Tropical Dry Zone- From the both sides of the equator, it extends from 25^0 to 30^0. The rain fall is very little. Hot deserts are seen in this zone. Days are hot and nights are cold.

(d) Winter Rain Zone- The name is given according to the rainfall in the area. Here summer is hot and dry and winter is wet and cold. This zone extends 40^0 north and south of the equator. This type of climate is found in California and Italy. It is called Mediterranean climate. Here trees are hard and woody.

(e) Temperate Zone- There is four kinds of temperate zones, i.e., warm temperate, typical temperate, arid temperate and boreal or cold temperate climate.

(f) The Arctic Zone- In winter season sun is not seen. The duration of summer is very short and sun light is cold.

Therefore, climate has important role in determining the environment. But climate depends on various factors; such as precipitation, temperature, humidity, wind and light.

PRECIPITATION:

Precipitation means all sorts of moisture that comes to earth in any form. This is possible when water vapours are condensed at high altitude. Water has various forms; such as snow, dew, hail etc. But water is very much essential for support of life. Due to sweating, when our bodies lose 2% of water, we feel thirsty; when 5% is lost, our tongues dry, mouth and skin shrink. But when 10 to 15% is lost, dehydration starts and becomes fatal. Water keeps the temperature of the living body constant. As water has high surface tension, it becomes a very good solvent for all biological purposes. When water becomes ice, it loses its density and floats on water. It has a good ecological value. The aquatic animals of cold region survive in under- water and come out when ice melts due to climatic condition.

Animals depend on water for their growth and other organic activities. The average rainfall in India is 117 cm. which is the highest in the world. But the parity of rainfall is not equal all over the country. Jaisalmir receives only 20 cm, Cherapunji 1100 cm, Assam 250 cm, Mumbai 200 cm, being in windward side, but Pune 75 cm, remaining in leeward side annually. So, vegetations and plantations grow accordingly, depending on rainfall. Likewise, in aquatic Eco-system, hydrophytes and aquatic animals grow in varying manner. Desert eco-system also grows according to the availability of water in that area.

TEMPERATURE:

It is a well known fact that, the physical activities of organisms depend on temperature. The rule of Vant Hoff says that in every 10^0c rise in temperature, the biochemical reactions double[4]. Animals can be classified into two types depending on the tolerance of the temperature. Most of the biochemical activities are done within 4^0 to 45^0c. The animals that can bear more temperature are called eurythermal and those who can bear low temperature are called stenothermal.

Not only animals, but also plants show variations due to change of temperature. Transpiration and respiration increase with the increase of temperature. Plants increase the nitrogen accumulation and carbohydrate reserve with the decrease of temperature. When the decrease of temperature is gradual, plants use more energy to receive nutrients and water from soil. The water absorbing capacity of plant is good when the temperature is within 25^0 to 30^0c. Along with the variation of temperature, we see different kind of changed behavior in animal kingdom; i.e.; frogs undergo hibernation, amphibians burrow in the soil, birds migrate from European and Russian climate to India during winter and return at the beginning of the spring etc.

In desert area, plants are also designed in such a way to avoid excessive heat of the sun. In order to get minimum heat, leaves are vertical and color is whitish or grey-green covering a thick waxy layer. So in distribution of organism, precipitation and temperature have important role.

---

4. Ibid-Page-191.

LIGHT:

Light has a pivotal position in metabolism of plants and animals. Excessive light impairs the flower holding capacity of plants, causes dehydration, minimizes photosynthesis etc. In animal kingdom, it affects the eye size and locomotion of animals. In oceans, those animals that live in 500 mt depth or more, their eye size is bigger than those, who live in the surface water.

BIOLUMINESCENCE:

Light, produced biologically by the animals is called bioluminescence. Animals have two organic compounds; luciferin and luciferase. When they are secreted by light gland and react with oxygen in water medium, light is produced. It is very much helpful in prey-predator relationship; i.e.; to catch the prey.

OXYGEN AND CARBON-DI-OXIDE AS ENVIRONMENTAL FACTOR:

Oxygen is the life of living beings. 21% of oxygen is found in free air and 4 to 10 ml per liter. In high altitude oxygen is very much scarce. Aquatic animals use dissolved oxygen for their respiration and metabolism. Fishes use gills for their respiration, some water animals also use their skin for respiration. Plants use carbon-di-oxide for photosynthesis. The rate of carbon-di-oxide in free air is 0.03%. When it is mixed with water, the respiratory activities of aquatic animals are influenced. Generally, fishes prefer stream water where carbon concentration is less.

PH AS AN ENVIRONMENTAL FACTOR:

It is the measuring unit of testing the purity of water. Here 0 to 14 is marked. The PH of distilled water is 7. From 7 to 14 is considered as alkaline and less than 7 to 0 is acidic. The distribution of living beings varies according to PH of water. The PH of estuarine water and sea water varies. The PH of estuarine water varies between 7.3 and 9, sea water between 8 and 9. So the role of PH is very much important in distribution of organisms in aquatic Eco-system. Discharge of effluents to the water bodies is changing the PH of water by creating threat to the fish and other aquatic organisms.

TOPOGRAPHIC FACTORS:

So many things like altitude, direction of mountain chains, slope, climatic conditions, distribution of organisms etc. constitute topographic factors. We all know that the air pressure gradually decreases, when we go up. When we go up 100 to 270 mt from the sea level, 1^0c of temperature falls down and the down fall becomes more when we cross beyond 1500 mt. It is more rapid in leeward side than wind ward side. Wind ward side also gets more rainfall. Mountain slopes play an important role in receiving rainfall, wind, velocity, solar radiation, temperature and above all on climatic condition of that area. The east side of the mountain is cooler than west side. Because it gets cool morning air, dew etc. But west side is already heated up. So environment is formed accordingly along with vegetation, soil erosion, rainfall and distribution of organisms. Mountains control the movement of the wind containing water vapour and rain near it. The opposite side becomes arid. Gangetic delta and north eastern region gets sufficient rain, so agriculture, vegetation and distribution of animals are more than other parts of our country.

EDAPHIC FACTORS (SOIL):

Soil is formed due to biological, chemical and physical weathering on rock. There are different natural processes, like heavy wind action breaks rocks, freezing water in rocky places expand and break rocks, flowing water grinds pebbles which form soil. Different types of acidic actions break rocks into silica, clay and inorganic salts etc. In the process of biological weathering different things are decomposed, acidic substances are produced and thereby break rocks.

Soil is divided into four types; A, B, C and D. Top soil is 'A' type of soil where different biological activities are done. It is enriched with organic compounds and color is black. 'B' type of soil is sub-soil and its color is light brown. Here biological activity is less than 'A' type of soil and contains clay and water. In 'C' type of soil; we find some long roots of big trees. But in 'D' type of soil, rocks are found.

The nature of soil is known according to the water holding capacity, aeration and presence of different types of particles in it. Different types of vegetation and agriculture are seen according to the quality of the soil. So soil can be sandy, silty, clayey and loamy. Sandy soil is heated quickly and water holding capacity is low, we see cactus in this soil. But in silty soil, the presence of humus is plenty, water holding capacity is also good and aeration is good also. Different types of plants grow here. In clayey soil, aeration is not good. Though water holding capacity is good, but water logging problem is found in rainy season. Soil is cracked during drought period. The growth of plant is not good. But loamy soil is the best type of soil. It contains sand, silt, clay and humus which is very much helpful for growth of plants.

Different types of micro and macro elements are present in the soil, which are used as food for microorganisms and soil animals and also increase the fertility. Macro elements are viz.; carbon, hydrogen, oxygen, calcium, nitrogen, phosphorous and sulphur etc. So also iron, manganese, zinc, boron etc. are microelements. Plants grow by taking water, nutrients and minerals from the soil through their roots. Distributions of plants are determined according to the presence of these items in the soil. Invertebrates and soil organisms are found more, where it is wet. We find more CO_2 in soil than atmosphere but equal amount of nitrogen and oxygen. It is a great edaphic factor, which determines the vegetation. Temperature of soil is a deciding factor of vegetation as well as soil organisms. Soil organisms include so many things, i.e.; bacteria, fungi, algae, protozoa, worms, earthworms, insects etc. Some mammals and birds also use soil. Earthworms burrow in two meters of soil and leave worm casts, filled with nitrogen and mineral substances, which are useful for plants growth. Not only this, they also facilitate aeration, water holding capacity by moving inside the soil.

BIOTIC FACTORS:

We find different types of interaction among species or individual in Eco-system, where some may gain or some may loss.

Symbiosis

It is a type of Eco-system, where different species interact among themselves for their benefit. It is seen that, certain fungi grow in the tree root; they also provide water and minerals to the tree in turn which they receive from the soil. In some sea animals, it is also found. The sea animals provide carbon-di-oxide as respiratory byproduct to the dianoflagellates containing chlorophyll for photosynthesis, which in turn produce growth stimulating factors for marine plants. So both the symbionts are benefitted.

Commensalism

Here one species is benefitted without causing any harm to the other. The other species is either benefitted or neutral. We see that one kind of crab dwells in the burrow of shrimp without causing any harm to it. Both the species are called commensalism.

Parasitism

In this system, one is benefitted and other is harmed. The species, that is benefitted is called parasite and the harmed species is called host.

Charles Elton said that "a parasite lives on host's income, while a predator lives on host's capital".[5] Some parasites reside on host's body either permanently or temporarily to satisfy their needs, i.e.; mosquitoes, louses, bugs etc. Vines take the support of host trees to climb up but they do not take food from it. Some kind of plants are carnivores, like pitcher plant of Assam, they trap the insect and take their enzymes, because their leaves are so designed.

GRAZING AS BIOTIC FACTOR

Its role is very much important in growth and structure of forests and grasslands. Due to excessive grazing, soil is eroded for runoff water. Ambasht said in 1974 that" in grazing seeds are consumed in large scale due to high nutritive value resulting poor successive crops.[6] Here herbivore animals are not only responsible, human beings also cut woods, use seed and fruits which cause loss of forest..

---

5. Ibid-Page-210.
6. Ibid-Page-212

FOREST FIRES

Lightening, rubbing of trees and human action cause forest fire. Human beings put fire for various purposes, viz., agriculture, habitat, construction of roads and establishment of industries and sometimes due to carelessness also. Forest fire is of three types; (i) Ground fire (ii) Surface fire (iii) Crown fire.

Ground fire can burn the litters and organic things on the ground. It burns everything excepting those trees whose barks are thick.

Surface fire burns the fallen leaves along with ground flora, which helps the growth of trees.

But crown fire catches the tree tops and jumps from crown to crown and destroys the entire forest. It is found in dense forest, where trees are adjacent to each other. Forest fire helps the growth of the trees by producing ashes by burning litters and organic compounds of the soil. It burns the harmful bacteria and fungi and helps the tree growth. So it is good, if it is done in controlled manner. In Odisha, our people burn forests in summer to get good Kendu leaves to be used for Bidi.

Community Ecology:

Different species live in a community by sharing a common habitat, resources and interact themselves. But in biotic community; animals, plants and microorganisms live together by depending on the environmental condition of the habitat. In a biotic community three classes are found, i.e., producer, consumer and decomposer. Every species has its own boundary, sometimes they cross it due to high rate of energy. This transitional system is called ecotone. If human beings will settle in forest, then forest will be reduced for different human purposes like housing, agriculture, roads and communication etc. So this human settlement will create ecotonal community.

Like ecotonal community, there is another word which is called biomes or biochore. It is a big area of same vegetation in different climatic condition. The nature of biome is determined according to rainfall, humidity, temperature, altitude and latitude. The diversity of community is determined on the basis of environment, light and water etc. The development of community depends on organisms. Two types of community are found; climax community and seral community. The community, which is at the mature stage is called climax community. The community, which is at the successional stage, is called seral community. Again succession is of two types; Primary and Secondary succession. In Primary succession, the seeds are brought from a different place and started in a new place. Germination is done and new off spring is sprouted. Then they increase their population. The first plants are called pioneers. The survival condition in primary succession is not good. In Secondary succession, the survival condition is good, because the land is full with nutrients. The abandoned crop land and cut over forest land come under this category. Here soil is full with organic and inorganic nutrients.

Population Ecology:

Population is that, a group of living beings live in a particular area and breed among themselves. Various things, such as natality, mortality, growth, age, density and distribution come under it. There is physiological limitation to natality, but no ecological limitation. Food and living places are limited. Environment has great effect on natality. If one female can be replaced by another, nataliy would be decreased. The child bearing capacity of a female is called biotic potential. So also mortality has great impact on population. People die due to various reasons; still there is a mortality age in every community. In each eco-system there is limited space and funds to sustain a fixed number of individuals. There are so many factors responsible for varying of population, like natality, mortality, immigration, emigration, change of environment, prey predator and producer-consumer relationship.

We mark certain positive and negative values during interaction among species. We find positive value in symbiosis and commensalism, but negative value in prey-predator relationship. Prey-predator relationship is a natural process of population control method. In 1968 America Government used sterilizing method to control the population of Culex mosquitoes by using the sterilizing agent theotepa.[7]

---

7. Kormondy Edward J, 'Concepts Of Ecology', 4th Edition, PHI Learning Pvt. Ltd. Page-237

In another instance, in 1907, the Kaibab plateau deer population in Grand Canyons in Arizona of USA was 4000 in number. So in order to save the deer population, all predators were killed by the government. But in 1918 when it was calculated, the deer population has increased up to 40,000, in 1924 it became 1, 00,000. So due to lack of food, 60% died in successive winters. Leopold has undertaken Kaibab study and suggested that predation is important factor in population control. On the other hand, if two species will depend on same resource then both will remain in critical position.

Another kind of competition the animals face within the same species for food, space and reproduction is called intra-specific. Some animals quit the society due to density of population and scarcity of food.

Only aggressive animals become successful in reproduction. The off springs those who come out become aggressive and quarrel among themselves for resources, space and reproduction and ultimately they are destroyed. Certain other factors which restrict population, like weather, water currents, chemical and other environmental factors. Bhopal gas tragedy, Super cyclone of 1999 in Odisha Coast, Tsunami in Indonesia and Japan are the burning example of it. In ancient time, cholera and small pox had smashed away villages after villages. For human birth control, different medical devices are invented. Infant mortality is checked due to proper medical care. Many developed countries have controlled the population explosion by applying various methods.

Natural Resource Ecology:

It satisfies human requirement in definite space and time. Natural resource is of two types: Non-renewable resources and renewable resources.

Non-renewable resources

There are different types of non-renewable resources, which are discussed below.

Mineral Resources

Which are obtained from mines is called mineral resources, viz.; iron, coal, aluminum, petrol etc. They are processed in the factory to get solid metal. These factories use coal, petrol or natural gas as energy. The rank of iron is first among natural resources. It is estimated that out of total iron deposited in the world, one fourth is deposited in our country, especially in Odisha and Jharkhand. Not only India is rich in iron but in 35 types of minerals. Oil is deposited in different river valleys, deltas and off shore sites. In South-West Asia, 60% of oil is reserved. Russia is the largest oil exporting country, but America is the largest oil consuming country which has only 7% of oil deposit.

Land Resources

It is the resource of resources. Our country has an area of 32, 88,000 km^2 which is about 2.4% of the world. Its population is more than 120 crores and remaining under great pressure of population. The pressure is gradually increasing day by day due to more industrialization and urbanization by giving rise to slum areas, which are hell in the earth. Therefore, land should be specified for different purposes like agriculture, forestry, grazing, water bodies, human settlement, road and industries for the healthy management of the resource.

Soil Resources

As each and every human action is done on the soil, it is the import resource. But formation of soil is very slow; it takes one thousand year for formation of one inch of soil. It can be eroded easily within a few span of time. Soil is formed by different ways, combined action of parent rock material, weathering process, climate, vegetation and decomposition etc. Plants are grown on it and provide food, fodder, fuel, medicine, fiber to the animal kingdom. It also catches water which is the life support of all biotic bodies.

Oceanic Resources

Different types of non-renewable resources are found at the depth of 4000mt - 5000mt below the sea level, such as gold, platinum, cobalt, copper, nickel, manganese, iron etc.

Renewable Resources

We find different types of renewable resources in the world; such as water resources, energy resources, agriculture range lands, forest resources, wild life and agriculture etc. which are discussed below.

Water Resources

Water is used by human beings for different purposes, like irrigation, production of electricity, industrial use, navigation and household use etc. Though there are plenty of water on earth, but only 2.7% of water is fresh water and out of that 77.2% is permanently frozen, 22.4% is ground water, 0.35% is in lakes and wet lands and 0.01% in rivers and streams. The underground water is called as aquifers. It is gradually diminishing due to use in agriculture and industrial purposes. It is estimated that a human being needs 2.7 liters per day and 1000 liters annually. It is seen that the distribution of fresh water is not in equal proportion all over the world. So in order to avoid water crisis, different measures should be taken to check the runoff water. In our country, we use only 10% of the total rainfall and the rest water run into the sea. We get 105 cm to 117cm of average rainfall per year, but now facing drought due to failure of monsoon, deforestation etc. Even the ground water level is going down and down due to excessive use of it.

Energy Resources:

It is of two types; renewable and non-renewable. Non-renewable energy is mostly used. Coal, gas, mineral oil come under it. Now-a-days petrol satisfies 45%, coal 25%, natural gas 20% of our total energy requirement. But by looking the scarcity of non-renewable energy, we are turning our face towards renewable energy, which are discussed here.

(i) Solar Energy

It is the main source of energy. Life on the earth is possible, only due to the presence of solar energy. It is the energy of energies. Coal, fossil fuel and gas are obtained when the forests are buried due to earth quake and under great pressure and temperature these were formed. Therefore, solar energy is the chief source of energy. Now technologies have developed to collect it through photocell, it has no side effect. So it is eco-friendly.

(ii) Wind Energy

Now-a-days technologies are developed to use wind energy for production of electricity, which can be used in small towns or villages. In Holland, the wide application of wind energy is seen in various fields of daily needs.

(iii) Wave and Tidal Energy

In estuarine area where river meets the sea, it is seen in violent form. We have vast coast line and all the rivers meet the sea in different coast lines, to which we can use best of it.

(iv) Geo Thermal Energy

It is the energy got from hot springs. There are 46 hot springs having more than 150^0c temperature in the different parts of our country, which can be used for production of electricity.

(v) Biogas

The gas arising out of dung of cows and cattle are called biogas. The people of our country foster different kind of domestic animals produce, huge amount of dung, which can be best use for production of bio energy and can be utilized in cooking and other household activities. Their left slurry can be used as manure in agriculture.

(vi) Biomass Energy

There are two types of biomass energy; non-renewable and renewable. Coal, fossil fuel and gas are non-renewable, but wood, cow dung are renewable. People in village areas use wood, cow dung for cooking which solve some part of energy crisis.

(vii) Atomic Energy

Atomic energy is produced in nuclear plants. Reactors are used to produce heat, which is used for production of steam. It is a great astonishment that one kg of natural uranium can generate energy, which is equal to 35,000 kg of coal. So it is an important source of energy. But on the other hand, it's devastating capacity is so pervasive that it can cause great loss from generation to generation. Recently, Fukusima accident of Japan has opened the eyes of the world. So people are protesting against installation of nuclear plants.

FOREST AS RENEWABLE RESOURCE:

Forests contain different types of flora and fauna and treated as natural eco-system. Previously, one third of total land was covered with forest, but it is diminishing day by day due to human action. Forests regulate climatic condition, protect soil erosion and satisfy our needs by providing food, fodder, fuel wood, timber, fiber, herbal medicine and other raw materials. It is a good habitat of wild life, which is the wealth of the nation. Forests have three functions; viz., productive, regulative and protective. In productive function, it transforms the solar energy into plant biomass and herbivore animals take it through their food. In regulative function, forests absorb carbon-di-oxide and release oxygen. In protective function, it protects the soil erosion from heavy rainfall and wind action. But every year we are losing 1.5 million hectors of forest per annum in our country and 4.5 millions of forest is already gone. Forests are destroyed due to various reasons, which are well known to everybody.

Social Forestry:

When deforestation brought so many problems for the earth, government felt the necessity of plantation and implemented various schemes accordingly. Plantations are made in government and private land, Panchayatas, road side, canal bank, along rail lines etc. Fast growing and commercially valuable plantations are taken into account to meet the necessity. Local people are involved in this work with a view to give good income to farmers. Acasia, Teak, Subabul are chosen which have high economic value as well as fast growing. But afterwards the aim of social forestry is modified and includes annual agricultural crop, farm animals and plantation of woody plants in order to maximize the output on sustainable basis. Today social forestry is called agro forestry.

Wild Life:

Living beings that stay in forests are treated as wild lives. Now-a-days their numbers are decreasing day by day with extinction of species. It is calculated that 88 species of birds and 100 species of mammals have become extinct during last two thousand years. One bird and one mammal species are under extinction every year. We have a great biodiversity, where 500 species of mammals, 1200 species of birds, 1500 flowering plants and 3500 non-flowering plants are found in India. Various types of wild animals live in our forests that provide meat, skin, horn, tusk for experimental research and recreational purpose. The International Union for Conservation of Nature (IUCN) has published two volumes of book regarding those animals which are in danger of extinction. The book is called "Red Data Book" which shows that 321 bird species and 277 mammals are endangered .[8]

There are three causes, which are very much important for this work. (i) Destruction of habitat (ii) Poaching by human being, for pleasure, meat, skin etc. (iii) Competition from domestic animals and transmission of diseases from them. Government is taking different measures for their protection by preparing various projects like crocodile project, Tiger project, bird sanctuary, reserve forests etc.

Range Lands:

These are very much vital for fostering of cows and cattle, as they provide us milk, meat, skin, horn etc. So we have to maintain forage production by regulating the frequency and intensity of grazing. To maintain rangelands, different points will be taken into account; like stock level, deferred grazing, fire. The stock of cows and cattle should be sold off during drought period. To get grown grass, range lands should be altered by grazing and non-grazing period. Fire work should be done during hot season

Agriculture:

In primitive period, agriculture was made on transitional basis. But now the scenario is changed. Different types of crops are produced in the same field annually, to meet the requirement of large population. So it has become now a man made eco-system.

Three things are essential for plants; soil, sun light and water. Plants use 45% of sunlight during 12 hours of day for photosynthesis. Our Indian agriculture mostly depends on monsoon rainfall. Pesticides and fertilizers are used in low amount. But to meet the growing demands of large population, green revolution was started.

---

8. Sharma P.D. 'Ecology and Environment', 10th Edition, 2009, Page-301

2 to 3 crops are produced in irrigated lands. Different types of cereals, pulses, oil seeds, fiber and fruits are produced depending on soil and climate. Rain fall is the deciding factor in production of crops. In Rajastan, Gujurat and Haryana, the rain fall is 10-70 cm per annum. Soil is sandy or alkaline or saline. But cows and buffaloes are fostered plentily, their dung are used in the fields and millet, beans, mung beans like low rain fed crops are produced. Sheep can be fostered to get hair for making woolen products. In hilly areas, slash and burn cultivation should be checked, as they help for soil erosion. Trees should be planted in the upper region of the hill. In the middle portion, fruit trees should be planted and in lower region certain crops which require low amount of water to be grown. Water should be logged by making earthen dams to be used during drought and for wild animals.

Aquaculture:

Rice with fish curry is a very good meal for people of coastal belt. Fish is also a proteinous food. We not only eat fish, but also crab, prawn, mollusces etc. We get 40% animal protein from fish. Fishes feed phytoplankton, zooplankton, weeds and other organic matters. We adopt two types of fisheries; capture fishery and culture fishery. Capture fishery is adopted in vast area like ocean and lakes. But culture fishery is found in artificial ponds. There is 56,000 km coast line in India, which provide better facility for fishing. Marine resources are innumerable in nature, such as; Bay of Bengal, Arabian Sea, Indian Ocean, numerous gulfs, coral reefs, mangrove swamp, estuarine of rivers, back waters, lagoon, brackish water lake like Chilika from where we can get plenty of aquatic animals. Apart from these, there are so many rivers, ponds, dams, irrigation channels where culture fishery can be promoted.

Today culture fishery is grown by using animal excreta like cow dung, domestic and agricultural waste and fertilizers. There are different types of ponds culture. Such as;

(i) Monoculture

In this system, only one type of species is developed.

(ii) Polyculture

Here different types of fishes are grown, like milk fishes with prawn, Chinese carp with Indian carp etc.

(iii) Intensive rearing pond culture

Here density of stock is heavy and rich in artificial food. So the quality of water is controlled accordingly. In such type of culture usually cat fishes, Eels and Trout are grown.

(iv) Extensive rearing ponds

Here the density of stock is low, so food supply is natural.

(v) Semi intensive rearing ponds

Food is supplied in both natural and artificial process. Fertilizers are added to increase food production.

Conservation and Resource Management:

The root word of conservation is conservare, a latin word which means guarding together. Con means together and servare means to guard. It's implicative meaning is that, non-renewable resources should be used in controlled manner and protected from wastage in order to save the environment. Charles Elton defined conservation as a "wise principle of co-existence between man and nature, even it has to be a modified kind of man modified kind of nature."[9]

Craggy said that "biological conservation is concerned with the maintenance of natural systems and where possible with their utilization either directly or by way of information obtained from their study, for the long term benefit of mankind."[10] This idea is very much befitting for the safety of the vulnerable species and proper management of natural resources. We can accept different steps for conservation of mineral resources; such as,

- New technologies should be developed to reduce and recover resources from waste.

---

9. Dash M.C. 'Fundamentals of Ecology', 2nd Edition, 2004, Tata Mc Grow-Hill Companies, Page-332.
10. Ibid-Page-332.

- We have to recycle metals.
- The tendency to use pure metals should be replaced by the use alloy metals by adopting new technology.
- Continuous efforts should be made to search the substitute of fossil fuel.
- The use of precious metals should be replaced by artificial products.
- Reclamation of mining areas.
- Strive to explore new oil mines.
- To regulate the use of mineral resources, a clear picture on its availability and expenditure should be maintained.

MANAGEMENT OF WATER RESOURCES, FORESTS AND WILD LIFE:

The major portion of earth is covered with water, but those are not usable. The usable water is very much scarce and we are making it more and more scarce due to mismanagement. We can use municipal water in industry and agriculture by filtering it. Run off rain water can be stored in ponds, ditches and trenches for future use. New technologies should be developed to desalinate the sea water for domestic and industrial use. This process is adopted in Bhavanagar of Gujrat.

Forests also play an important role for the existence of the earth. So following measures should be taken for protection of forest.

- Rampant cutting of trees should be stopped.
- Plantations should be made to create new forests.
- Forest fire should be forbidden strictly.
- For the growth of forests, grazing should be controlled.
- Proper care should be taken for growth of plants.
- Careful and scientific harvests of forest products.
- For production of fuel wood, timber, fodder etc. social forestry should be adopted.
- The most important thing is that, people should be made conscious for preservation of forests and wild lives.

Chipco and Apico movement were made successful and got world support for preservation of forest and Eco-system due to public consciousness. Preservation of forests and restoration of wild habitats go side by side. If forests will be preserved wild animals will be saved.

RECYCLING OF RESOURCES AND WASTES:

Due to excessive industrial action, waste materials are produced in gigantic manner, so technology should be developed to recycle these things. We use cow and cattle dung for production of biogas. So also we use solid wastes for production of earthworm tissue, worm cast. Again we can use these worm casts in production of mushroom, vegetable and crop. Earthworm tissue can be used in poultry, pig and fish farms as feed. Metal wastes can be used in cement in the form of powder. The metal got from machines, ships and aero planes can be recycled. Recycled municipal water can be used in agricultural fields.

Earthworms play an important role in degradation of organic wastes, which can be used in various fields; such as agriculture, aquaculture and poultry. It has a very good capacity to balance the nutrient flux in Eco-system. The organic manure can be used in various ways.

- It acts as a substitute of chemical fertilizer.
- The quality of soil is improved as it gets humus from it.
- The humus is increased; water retention capacity is also increased.
- Soil pollution is controlled due to use of organic manure.

To-day, such type of technology is developed, which is called vermitechnology in both developing and developed countries, as it helps for recovery of resources and waste management.

POLLUTION ECOLOGY:

Pollution is undesirable change in nature, which brings harms to living organisms. It changes the physical, chemical and biological character of air, water and soil. Pollution takes place due to two types of substances, i.e.; persistent and non-persistent. Persistent substances are those, which are not easily degradable. Different kinds of plastics, pesticides, and nuclear wastes come under this group. Whereas non-persistent substances are easily degradable, like house hold garbages, agricultural wastes etc. Now we will discuss how different kinds of pollutions are endangering our lives in various ways.

Air Pollution

Life, without air is inconceivable as all the living beings breath air. It has certain weight and pressure. It is high at sea level and low at top level. The ingredients of common air vary in proportion, like 78.09% of nitrogen, 20.94% of oxygen, 0.93% of argon, 0.032% of carbon-di-oxide etc. But air pollution takes place, when any kind of alternation brought in the air with the introduction of harmful gases or particles. We are using vehicles, aircrafts, factory chimneys, fire for house hold cooking, which are causing pollution. Different kinds of air pollution are as follows:-

- Different kinds of fossil fuels are used for various purposes, which produce carbon-di-oxide, carbon monoxide, sulphur dioxide, nitrogen oxide and smoke.
- Sulphur oxide, fluorine and particulate mattes are produced in large scale in fertilizer plants.
- Thermal plants pollute air by producing fly ash and sulphur-di-oxide pollutants.
- Various types of pollutants come out of textile industry, such as; cotton dust, nitrogen oxide, chlorine, naphtha vapours, smoke and sulphur-di-oxide.
- Chlorine gas is emitted by chemical and pesticide plants.
- Iron and steel industries produce carbon monoxide, carbon-di-oxide, sulphur-di-oxide, fluorine and dust.
- During the period of decomposition of organic matters, foul smelling gases are produced and pollute air.

Certain Primary Air Pollutants

Air is polluted in various ways, which is not possible to discuss all the factors responsible for it. Only some primary pollutants are discussed here under.

Carbon Monoxide

Automobiles mainly produce this gas. It is also produced from furnaces, stoves, open fires, factories, power plants etc. This gas is destroyed by fungi and other higher plants, which depends on the availability of solar energy. When this gas is mixed with hemoglobin of blood, it affects the respiratory activity and metabolism by reducing oxygen carrying capacity. Headache, blurred vision and acute toxicity are caused by it. It has more affinity than oxygen to mix with blood. So it controls the supply of oxygen from lungs to different parts of the body.

Carbon-di-oxide

During the time of photosynthesis, plants need this gas. Excessive use of fossil fuel and coal cause massive production of this gas, which disturbs the heat balance of the earth. The sun light which falls on earth, some part is reflected back to sky and some to the earth. In continuation of this process, earth is heated. This is called green house effect. If CO_2 will increase in this rate, the temperature will be doubled in 100 years. Ice of the Polar Regions will be melted. Most of the important cities like New York, London, Tokyo, and Kolkata etc. will be submerged. Agriculture, climate and rainfall will be affected.

Sulphur-di-oxide

It is also produced from the use of fossil fuel and coal. There are three types of sulphur dioxide; SO_2, H_2S and sulphur particles. When SO_2 is oxidized, it becomes SO_4 and comes back to earth through rain. It can be easily dissolved in water, so it can easily enter into soil and vegetation. It controls the carbon dioxide receiving capacity of plants. So leaves are destroyed and yield is reduced. It creates respiratory problem of the animals. Paper becomes yellow and loses when comes in contact with it. SO_2 affects iron and steel, lime stone, marble and leather etc.

Hydrocarbons

It is also produced from automobiles. There are different kinds of hydrocarbon; like methane, ethane, propane, toluene, m-xylene etc. When these gases are mixed with water, secondary pollutants are produced. Methane produced from organic wastes also.

Nitrogen Oxide

It is found in automobile exhausts. It can damage the plants in low concentration, but affects the respiratory system of mammals in high concentration.

Photochemical Smog

It is also produced from automobile exhausts. The color of this gas is yellowish brown. It is produced, when hydrocarbon and nitrogen oxide are mixed with each other in the presence of sun light. Living beings as well as environment are affected. It brings various types of harms; like head ache, eye irritation, sore irritation, respiratory irritation and reduction of plant growth etc.

Acid Rain

When sulphur-di-oxide and nitrogen oxide reacts with water in atmosphere, sulphuric acid and nitric acid are produced and come down to the earth through rainfall, which is called acid rain. The acidity of water is measured on PH scale, where 0 to 14 is marked. If PH value of water is less than 7, it is acidic and more than 7, it is alkaline. It is seen that natural rain fall is slightly acidic, the PH is 6.

The causes of nitrogen oxide and sulphur-di-oxide are forest fires, volcanic eruption, electric generation plants, smelting plants and oil refineries etc. The oil refinery of Mathura is creating problem for Taj Mahal, which is only 40 km from it. Marble monuments and lime stone works are badly affected. When soil and water become acidic, it dissolves aluminum with it and affects the fishes by damaging their gills. The leaves of trees become brown and yellow. Aquatic animals are badly affected, because it ruins planktons and fishes.

Aerosols

These are particles, which create problem for aircrafts. Dust particles absorb other substances like hydrocarbon, sulphur, nitrogen oxide and float on the air.

Depletion of Ozone Layer

It acts as an umbrella over earth by checking the ultraviolet rays of the sun, which can cause skin cancer for the animals. But a big hole was found in this layer over Antarctica. Therefore, Montreal protocol to Viena convention was held in 1985 and delegates from different countries reached at the decision to minimize the emission of chlorofluro carbon by 50% by the middle of 1998, which is the main cause of Ozone depletion.

There are certain other pollutant gases, whose minute quantity has good effect, but higher dose makes the water toxic and damage vegetation, i.e.; fluorine gas. So also chlorine gas damages the leaf stomata and cause breathing problems of animals.

DIFFERENT MEASURES TO CONTROL AIR POLLUTION:
- In order to control air pollution, modern equipments and smokeless fuels should be used in industry. Scrubbers, filters and precipitation systems should be taken to control industrial particulates.
- The emission of smoke can be checked by using biogas, smokeless chulas and solar cookers etc.
- Long chimneys should be fixed in coal using industries to disperse pollutants over a large area.
- It is known that coal produces sulphur-di-oxide. So it should be replaced by low sulphur containing coals or other energies.
- Anti pollutant devices should be fixed in vehicle to filter pollutants.
- Wood, cow dung or farm wastes should not be used in cooking as they produce too much smoke, rather electricity or biogas can be used.
- Public consciousness should be made to prevent pollution.
- Plantation should be done in different places, as they are good controller of pollution and provide us food, fodder, fiber, shelter and fuel etc.

WATER POLLUTION:

It means adding of various organic and inorganic substances in water by human action, which is harmful to living organisms. It is peculiar that, we are living in one fourth portion of the earth but polluting the three fourth portions of the water bodies. Water is polluted by different sources, which are discussed below.

Waste Materials

Dumping of waste materials near water bodies which mix with water afterwards, pollute the water and makes the aquatic Eco-system unsuitable for aquatic animals living there. In our country; ponds, rivers and lakes are almost polluted by sewage, agricultural wastes, and animal excreta by depleting oxygen and various diseases like cholera, typhoid, jaundice etc. are caused.

Eutrophication

The excessive storage of nutrients in water bodies is called eutrophication. Due to growth of algae and other organic matters in water bodies, water becomes green and more oxygen is needed for respiration of aquatic animals. As the water of agricultural fields is full with nutrients, they make the water bodies green, when mix with them.

Industrial Wastes

Industries are installed near the rivers or seas and dump the wastes containing different chemicals, acids, alkaline by the side of it, which causes serious problems for aquatic and terrestrial animals. In our daily life, we use detergent for cleaning our clothes. They mix with water and pollute the water bodies. Once in London the river Thames was covered with detergent foam and looked as if the river is full with white swans. Paper mill effluents cause serious problem for the aquatic animals. Steel industries, nuclear reactors, electric power plants require huge amount of water for their cooling process and leave the hot water into different water bodies, which deplete oxygen and harm the aquatic animals.

Pollution by Oil Spills

Sea routes are used for carrying of oils through large tankers from one place to other. During this period, when any kind of accident occurs, tanker is licked and oil mixes with water. Then it floats on the surface water and prevents atmospheric oxygen to mix with water and create respiratory problems for aquatic animals of that area.

Agrochemicals and Water Pollutants

In cultivation process, different types of fertilizers and pesticides are used in the field to get more yields. But after harvesting the residues of these particles remain in the fields, those are run off with water and the water becomes polluted. When animals take this water, they suffer or die. There are some pesticides, which are not biodegradable and long lasting like D.D.T, Aldrin and Dieldrin. They enter into food chain and stored in the intestine of animals which damage the fertility of animals and nervous system.

Water Pollution by Metals

There are different kinds of metals emitted by industrial operations, mix with water and pollute the water bodies. Here I want to discuss some metals of this type.

Mercury

When mercury is emitted by industries to the water bodies as waste material, it pollutes the water. The fishes, living in that area take it through their food like plankton and magnify it during different stages of metabolism, then enter into the human body and create different types of health hazards. Such type of problem was found in Japan in 1953. At that time, Minimata Bay was contaminated by methyl mercury and people taking sea food from that area, suffered from the disease which was called Minimata disease. The effect of the disease was numbness of limbs, lips and tongue, causing blurred vision, deafness and mental derangement. At that time, seventeen died and twenty three became disabled. Mercury pollution is caused by paper mills and chlorine producing industries. The Ib river of Odisha is polluted due to paper mill effluents of Brajarajnagar.

Lead

In European life style of Roman and Greek, the use of lead was found as it gives good taste to food and drink and retards decomposition. They prepare grape syrup in lead line pot to get good taste. But its excessive use gives rise to lassitude and infertility. Automobile exhausts which have lead contamination, when enter into body through respiratory function affect brain, peripheral nervous system, vomiting, diarrhea, loss of appetite etc.

Cadmium

It is stored in the earth's crust with other metals like zinc, lead-zinc and lead-copper-zinc ores. It is used for electroplating of metals. The waste materials of cadmium industries are found in the form of fume, dust, sludge and waste water etc. Some of the cadmium residues like cadmium fluoride and cadmium bromide are soluble in water, but cadmium sulphide and cadmium oxide are not soluble in water. Near about 2000 tones cadmium compounds are released to the environment every year. They are stored in different parts of the body like liver, kidney, testes, lungs, pancreas, and endocrine organ etc. and cause nausea, vomiting, headache, shock, chest pain, bronchial pneumonia and kidney problems etc.

Arsenic

Since the beginning of human civilization, it is used in various ways like; protection of food products, crops, for production of medicine by using in small proportion etc. However in different use, it enters into our body and when becomes acute, liver, kidney, nerve disorders gastro intestinal damages are found.

Mining Operations

Mining operations are gradually rising with the rise of human population. Due to lack of modern technological knowledge, we are adopting traditional method like open caste mines, for which water in that area is polluted and local people suffer from various water borne diseases.

River Pollution in India

People started living near the bank of the rivers due to high fertility of the soil. Accordingly township and industries have also grown, which are important factors responsible for pollution. As the municipal wastes and industrial wastes are dumped near the river, the water gets polluted. The oxidizing capacity of water becomes lower and lower due to excessive dumping. Aquatic and terrestrial animals, those who depend on it are mostly affected. Once we were using the water of river Ganga in rites and rituals as holy water and still it is continuing, but the water of Ganga has lost its purity due to excessive dumping of industrial wastes and municipal garbage, sewages and other human activities. So Government of India is preparing master plan to keep the river clean.

Control of Water Pollution

Now-a-days, it is very much important how to control water pollution, because life without water is not possible. There are three stages of water pollution, viz., primary, secondary and tertiary.

Under primary treatment, we use screens or sieves to separate solid objects. Then the water flows into settling tank, where the suspended materials are sludge down into the tank.

In secondary treatment, the organic materials are permitted to be decomposed by bacterial operation, being filtered through a bed of rocks. Air is bubbled to increase oxygen. Chlorine is used to kill the germs present in the effluent.

But in tertiary treatment, the effluents are absorbed through activated charcoal when these are passed through it after making secondary treatment.

We can use polluted water in agricultural fields after some primary treatment and it is less expensive also. But the most befitting method to control pollution is public consciousness.

SOIL POLLUTION:

Soil is used in various ways and also polluted in various ways. We use poison in agricultural fields, where harmful and beneficial organisms are killed. There are certain herbicides, whose particles remain in the soil and harm the wild life and human life. Earthworms which are eco friendly, they raise fertility of soil by decomposition, cycling of matters etc. badly affected insects, birds and amphibians those who take earthworms they are also affected.

Soil is also polluted, as human beings leaves excreta in the open fields. They mix with water and different diseases like cholera, typhoid affect people who use these water. It is found mostly in village areas. Because, people living in those areas, use river or pond water. World Health Organization gave a statistics that about 15 million children die before the age due to lack of sanitation. Apart from this, soils of industrial areas are polluted due to dumping of industrial wastes.

DRUG ABUSE:

Our young mass use drugs of various kinds and think it as the symbol of aristocracy at present. Even if some educated people use drugs without thinking about the consequences. Wyrobek and Bruce had proved in 1975 that these habits badly affect our sperm producing capacity.

NOISE POLLUTION:

Production of unpleasant sound is the cause of noise pollution. The measuring unit of sound is called decibel and we can hear from 0 to 130 decibel. Beyond this it can cause physical damage to our ear. It can be best understood, if we take different examples of sound. At the time of chewing gums 20 decibel, type machine 40 decibel, conversation 30 to 60, telephone sound 70, motor cycle 110 to 120, siren 130 to 150, jet plane produce 160 or more at the time of take off. Noise pollution is of three types, viz., industrial, traffic and community.

Industries create noise pollution, when these run with heavy machineries, which affect the workers engaged there. In towns and cities, automobiles always run creating different sounds. That is called traffic sound pollution. Community sound pollution is created in different social functions like marriage, political and religious gatherings. Hospitals and schools are sometimes affected by it.

Government has framed different rules to control noise pollution. Industries should be set up far from cities and towns. Aerodrome, highways should be away from the populated areas. A venue plantation should be started on both sides of the road. It can also reduce noise up to 20 decibel. Noise pollution can be checked, if public conscious can be roused.

POLLUTION FROM RADIATION:

The world has seen atomic explosion of Hirosima and Nagasaki and experienced different nuclear accidents, like Chernobil and Fukusima. So people are well aware about its consequences. Atomic radiation is of three kinds; alpha particles, beta particles and gamma rays. These rays can damage the tissues of living bodies being penetrated into it, though these are invisible. These rays can penetrate into body in various ways. These rays obstruct the activity of living cell and create free radicals which react with other compounds. The unit of measuring radiation is called 'Rad'. When a living body will be exposed to 1000 rad it will create cancer or sterility. The genetic makeup of body is changed due to radiation. So radiation damages cannot be repaired. It has bad effects on vegetation. But it has certain positive value also. It is used for diagnosis of certain diseases in controlled dose.

From the longevity aspect, the nuclear wastes last for long time, i.e.; for thousands of years. So the developed countries dump the nuclear wastes in the deep sea, keeping inside closed containers. Therefore, experiencing the devastating capacity, people are turning their eyes from nuclear power.

In order to study the cross cultural attitude of men towards environment and ecology, first of all; I feel it necessary to have a bird's eye view about environment and ecology which is purely descriptive. In the next chapter, I shall have a philosophical search into the attitude of modern men. So, mechanistic, humanistic and vitalistic models come to our discussion.

****

CHAPTER-II

METAPHYSICAL MODELS AND CONFLICTING CULTURAL PATTERNS: IT'S IMPACT ON ENVIRONMENT AND ECO-SYSTEM

METAPHYSICAL MODELS AND CONFLICTING CULTURAL PATTERNS: IT'S IMPACT ON ENVIRONMENT AND ECO-SYSTEM

In this chapter, I want to show how two metaphysical models i.e., realism (ultimately finding a way for materialistic attitude); and idealism (ultimately finding a way for spiritualistic attitude); have created two cultural patterns, which are conflicting with each other in a set up, failing to promote humanism, which keeps the interest of human beings intact, at the past, present and future.

REALISM:

As a matter of influence of Cartesian philosophy, two metaphysical entities, i.e., (i) Body (Matter), (ii) Mind (Spirit) became most predominant in the level of thought, which influenced a lot the social system. It is Descartes who first of all gave a clear picture of body and mind as two independent realities. Although Descartes wanted to keep philosophy on a sound and solid base by discovering the universal truth "Dubito cogito ergo sum"[1], yet, later on he discovered body and mind as two independent realities which in turn created the dualistic system of thought. Afterwards realistic (materialistic) philosophy found its own domain of explanation mostly in the thought of John Locke, separating it and with an opposing attitude towards idealism (spiritualism).

Locke's realism is, perhaps, an outcome of the Naive realism, which says that the matter is out there and real however Lockean realism is called the representative realism which runs like this;

- There exists a world of physical objects (trees, buildings, hills etc.)
- Statements about these objects can be known to be true through sense experience.
- These objects exist not only when they are being perceived but also when they are not perceived. They are independent of perception.
- By means of our senses, we perceive the physical world pretty much as it is. In the main, our claims to have knowledge of it are justified.

---
1. The Principles of Philosophy, Part-I, VII

- The sense-impressions we have of physical things are caused by those physical things themselves. For example, my experience of the chair is caused by the chair itself.[2]

Although realism is not exactly materialism, yet it paves a way and supports materialistic attitude which is technically called mechanistic theory with the emergence of and dominance of empirical philosophy, realistic philosophy; mechanistic view rose to its apex and viewed everything in terms of matter. The matter is all in all, centre of everything; even 'man' is the outcome of materialistic evolution. Man is nothing but a machine , a sophisticated machine which evolutes out of the matter which we see in the external world; according to mechanistic view there is nothing spiritual or there is no such thing as life force which can be seen in man. Man is simply a product of amalgamation of matter, or mixture of chemical components, this is what the empiricists, realists, materialists opine and this view, as a whole is called a mechanistic view.

This mechanistic attitude of man which considers man as a machine has tremendous effect on society, creating a vast cultural pattern with a strong hold on philosophical basis in the entire world; a mechanistic culture prevailed, as a result of which industrial revolution, economic growth, production, market value - these concepts became most predominant in the social sphere. Monetary gain, materialistic development became the be all and end all of human life. It's greater impact found its own way in the field of industrial revolution, where man started snatching away natural recourses and processing these in the mega factories and availing huge quantity of products, marketing these products in a global basis. The above said circle of materialistic greed created a society, where we find a cultural pattern; making a way for selfish gain, fulfillment of unlimited human greed. There is a saying "the world is enough for the need of cores of man but insufficient for the greed of one man".

---

2. An Introduction to Philosophical Analysis by John Hospers, Chapter-VIII

As mechanistic attitude, which turns a deaf ear to value system in man, and always gives importance on production and economic growth, a cultural pattern evolved out of it, where the awareness of environmental factor was thrown to abysmal depth. Natural recourses like minerals, forest products etc are exploited ruthlessly in large scale creating a great hazard to the environment and ecosystem. Deforestation is a common phenomenon now-a- days resulting in abnormal rain, global warming, acid rain etc. the most important thing in this cultural pattern is that man becomes more and more greedy losing all sorts of value system for which the entire environment and ecosystem may collapse like a house of cards. The world foresees the horror of destruction due to a pseudo cultural pattern, i.e., the mechanistic pattern of society.

IDEALISM:

The chief exponent of idealism is Berkeley. He strongly opposed the realism of John Locke, he believed that there was no basis whatever for holding a distinction between primary and secondary qualities. Secondary qualities, Berkeley argued, are inseparable from primary ones: if the object does not have the one, it cannot have the other. For example, colour and shape are inseparable. Consider any shape you please, such as you might draw on the blackboard or paint on a canvas. What can feel the shape except a colour? Shape, said Berkeley, is simply the boundary of a colour. You cannot even imagine a shape without a colour. Whatever its shape may be, it must be filled up by a colour. So if shape is a primary quality of object, then so also the colour is.

A thing may indeed, said Berkeley, appear to have different colours, depending on the conditions of the environment and internal condition of the observer. But if variability proves subjectivity-that is, if it proves that, object does not itself possess these qualities – then Locke's argument proves too much, for it applies not only argument proves too much, for it applies not only to colours and smells but to shapes and sizes and other so called primary qualities as well. A thing will appear to have different shapes, when viewed from different angles. The coin looks circular seen from above and elliptical from various angels. Things look different sizes, depending on one's distance from them.

Hence, shape and size are variable no less than colour and smell. Berkeley holds that there is no distinction between primary and secondary qualities; we may end up with physical qualities and sensible qualities.

Locke says that our sense experiences (ideas) of primary qualities are resemblances of these same qualities, as they really do characterize the object in the outside world. But our experiences of the secondary qualities don't resemble any quality in the object for there is in the object, no such quality: there is only the power to produce certain sense experiences of colour, sound etc. But now says Berkeley how could we ever know that our experiences of primary qualities resemble those qualities in the object itself? How could we ever compare them to discover this? We can speak of comparing only when we are in position to compare; we can compare two colours-for example, say that this one is lighter than that one –because we can see them both. But how can we compare our experience of shape or size with the allegedly real shape and size of the object, which is independently of our experiences? We could not possibly do so - it would in fact be logically impossible to do so, because to do so, we would have to be able to experience both of the items to be compared. And this, we can't do. We are acquainted only with our own experience, and we cannot experience anything other than them. An idea, said Berkeley, can resemble nothing but another idea. This means that sense experiences can be compared with other sense experiences. What would it even mean to say that the experience of size is like the objects real size? There is simply nothing that you could say about them. We can say many things about our sense experiences, but we can't say anything about the physical world "as it is in itself", that the sense experiences are supposed to resemble. Thus, on the question of whether the experiences we have of shape and size resemble the real shape and size of the object, Locke is condemned to a total and incurable scepticism.

This scepticism must extend to every property that the object "as it really is" is supposed to have, including it's causal properties. Since, we all can know our own experiences and Locke had admitted this for himself, when he said that "the mind has acquaintance with only its own ideas"- we can have no way of knowing whether our experiences resemble the object themselves or whether the objects cause the experience either. In order to solve this problem, Karl Pearson gives the analogy of telephone exchange. The mind is like a telephone exchange; you are the telephone operator, messages come into you along the wires from the outside world. You don't see the people who do the telephoning, you hear only the sounds of outer voice, you receive the incoming call via incoming wires and you connect them with their proper parties. But you yourself never get outside the exchange. In other words, an agent is within the sphere of sense experience or sense data, can react and other than the sense experience, it has no passage. Thus far, we have given a brief criticism of Berkeley against Locke's representative realism.

Although Berkeley's idealism never speaks of spiritualism, yet, this type of thought paves a way for spiritualism. Now let us see, what Berkeley wants to say in his contention. According to him, Locke had no reason for holding his view about the existence of physical world. Locke is committed to scepticism regarding a physical world; he can't know that it exists, even if it does; and he is inconsistent, because he assumes that it exists and makes claims concerning it, yet cuts himself off from the possibility of knowing it, which invalidates the arguments about physical objects and there qualities that he had given just before.

Berkeley is not denying that there are trees and books etc, but he is denying that there are any physical things in the sense of objects that exists independently of minds. In short, Berkeley is an idealist. All that exists are mind's and experience of mind's, but there are no independently existing physical objects to cause the experiences. The experience of objects is enough; object itself is an excess baggage. Berkeley did indeed believe that, there are chairs, but not that our chair experiences are caused by chairs- i.e., by physical objects existing outsiders and independently of us. Berkeley says, if by physical objects, you mean groups or complexes of sense experience, then they undoubtedly do exist-indeed, we are aware of them, every waking moment of our lives, since we are constantly fall in to ordered patterns or groups.

SPINOZA'S THEORY OF SUBSTANCE:

By substance Spinoza understands "that which exists in itself and is conceived by itself, i.e., that which does not need the conception of any other thing in order to be conceived".[3] From this he deduced the following conclusion as to the nature or character of substance.

Substance is its own cause (causa sui) if it were produced by any other cause, then it would depend upon that, and would, therefore cease to be substance.

Being the cause of itself it is absolutely independent and infinite. It is obvious from this that there can be only one substance, for if there were two or more, they would limit one another and would thereby cease to be independent and therefore to be substance. Hence, there can be one substance, which depends on nothing but on which everything else depends. For Spinoza, God alone is substance and substance is God; the two terms are thus synonymous. Here we find an indication to have spiritualistic attitude.

MONADS:

According to Leibnitz, the world of bodies is composed of an infinite number of dynamic units or immaterial, unextended, simple units of force or monads. What can we say about these monads? Leibnitz says that, these monads can be well conceived of, after an analogy to our own selves, we discover such a simple, in extended and immaterial force- unit as the monad in our own inner life. The soul is such a substance, and what is true of it will also be true of all monads. The reasoning by analogy, he interprets the monads as so many spiritual or psychic forces like our souls. Just as the human soul has the power of perception and conation, so also the monads are endowed with perception and desire or appetition. Thus all the monads of all stages, whether lower or higher, whether the lowest or highest, whether the most imperfect or perfect are entelechies or souls. That Leibnitz theory of monads indicates to a world of spiritual existences.

3."The Ethica" by Spinoza

Thus, there are two opposing points of view, known as mechanism and vitalism, mechanism emphasizing the continuity and likeness between living and non-living things and vitalism the discontinuity and difference. A very common way of distinguishing between mechanism and vitalism: according to vitalism, there is a special non material life force, or *elan vital* which is present in living things and not in non-living things. It's presence in living things explains the difference between the behavior of living things and that of non-living as said above, it appears that there are two metaphysical models conflicting with each other, i.e., two opposing characteristics. And we will find out that these conflicting metaphysical models have diversified the society into two conflicting cultural patterns, i.e., mechanistic and vitalistic.

SPIRITUALISM:

As is evident from the discussion so far conducted, a metaphysical model is most powerful to influence the pattern of the culture in the society. The vitalistic type of metaphysical model gave a boost to the spiritualistic pattern of society in which religious theories, ethical values of different schools developed rapidly. There were different schools of spiritualism in the society where different types of people participate to their own choice. The most important and significant common programme to be seen in different schools of spiritualism is that they promote ethical values which has longstanding repercussion on human civilization. Ethical values promote peaceful co-existence, normalcy in the society and the most important thing is that it has given a yeoman's service to combat environmental pollution and ecological imbalance. In the history of civilization, a remarkable factor to be found is that ethical value has turned heaven and earth to protect ecology and environment. In some of the religious scriptures, we find ritualistic element in the flora and fauna and also in the inanimate objects. May it be true or false but a greater impact of this thinking finds a massive influence to protect ecology and environment worldwide.

The greatest drawback of spiritualism is found when it reaches its apex. That's why there is a saying- spiritualism or faith pushed to its extreme defeats for its own sake. Spiritualism or faith should be inculcated to that extent where man can lead a normal and comfortable life in a harmonious society.

Being over powered by the spiritualistic sentiment, some religious followers have created havoc in the society, creating a great damage to environment and ecology. The most crucial problem, the world now beholds is that, there is rampant population explosion, which is the key factor of erosion of ecology and environment .The society at present is in earnest need to control population for the protection of environment and ecology, but some religious systems say some spiritualistic patterns and extreme spiritualistic pursuit; holds blind belief and say that family planning is against the will of God. Spiritualism having a powerful impact on men in the society, here gives a boost to population explosion indirectly giving the greatest damage to environment and ecology.

We find innumerable blind beliefs in the field of spiritualism, which is not possible to mention all. We are just mentioning a few instances for reference. In the history of civilization, we find that the great wars were fought due to spiritual conflict or religious conflict which caused a great destruction to the earth as a whole. The war between the Uddyans and Islam, the war between the Gods and Demons has caused a great destruction to the society causing environmental pollution and ecological imbalance. Even Nazism, though racial, is yet greatly influenced by spiritual factors which were the main cause for world war resulting in chemical and atomic weapons used in the huge scale, distempering ecology and environment. The application of nuclear weapons at Hiroshima and Nagasaki found great devastation, the effect of which is found now-a-days; lakhs of people are suffering from neuro diseases and cancer etc. in Japan.

Even in the ethical side, we also find some theories such as gross egoistic hedonism which promotes snatching away and damaging natural resources causing environment and ecological problems. Rigorism invites acute spiritualism, though not so harmful for environment and ecology, yet creates abnormalcy in the normal way of life. Perfectionism which is a synthesis in between rigorism and hedonism is coincident with humanism, which is most favorable for ecology and environment.

THE HARMONY:

There is no doubt that different people are situated in different levels of spiritual understanding. By their presence, Knowledge and Practical realization, some are able to create a cheerful atmosphere around them. Others, due to contrary emotional values, estimations and selfish interests, are inconsistent. At a lower spiritual level, emotion in relation to belief, principles and dignity are often flickering and imperfect. In that dimension, the standard of understanding, the surrounding is often deficient. The magnitude of feelings, being impaired by doubt, fear, and dissatisfaction, are unsteady. Having ideas based on faulty distinctions, they are often the source of all disturbances.

To reach a higher dimension, the seer has to give up the pride of old experiences, which is the basis of how he perceives things. He has to be aware that his flickering feelings, opinions and impressions are certainly not true statements. Those can change at any time. In the intellect, these misleading and deceptive thoughts are forcibly stored. To clear up all these misconceptions, new and fresh ideas have to be received from authorized sources. One has to investigate how to eliminate these erroneous percepts and accept factual considerations. He has to analyze all possible answers to each question followed by a systematic rejection of any undesirable lower misconception of life. To find out the truth, a new insight has to arise from observing ordinary events. Except for oneself, none can perceive and check one's further progress. To brush aside any undesirable concept, one has to be above the common spiritualist, who sees every day, many happenings without examining their inner truth.

On this spiritual dimension varieties exist. But they should not result in conflict. Each individual is related to God, the source of all harmony. As the spiritual is free from all kinds of sorrows, it is the key to all harmony. Just as the rising sun dispels darkness and brings auspiciousness everywhere, the spiritual drives away all conflicts. However, materialistic propensities are very difficult to harmonize. They are often tainted with strong emotions that are filled with hate and anger. On the spiritual level, unbearable happenings are ultimately seen as the will of the Supreme. For instance, Joseph and Jacob accepted all events as the will of God. Jacob knew that, out of jealousy, his other sons planned to kill his darling small child Joseph, but Jacob never showed any enmity to his envious sons. As a small child, his brothers threw Joseph in a well. He was sold as a slave. He was unjustly put in the prison for many years. Yet, when Joseph became a great personality in Egypt, he did not show any hostility to his brothers, who desperately needed his help. Rather he received them with great respect and humility. He supplied them with whatever they needed. Because of the goodness exhibited by Jacob and Joseph in handling such a sensitive matter, Jacobs's sons were greatly benefited. With the passing of time, their enviousness towards their brother increased gradually, Joseph faded away. Jacob could have brutally damned the act of his sons to kill Joseph, but by refraining from doing so, everything ended favourably.

The main ingredient of quarrel between men is self aggrandizement. Man wants to show his greatness to his family members, his community and his surroundings. To achieve such purpose one is ready to manipulate the message and everything around him. The tendency of self aggrandizement is dangerous. It is self destructive and detrimental to the principle of dedication to God. No healthy contribution can come from a proud boasting person. In such a person, God is not the centre, rather one's selfish interest is. Both Vedic literatures and the Koran aim to cut off all personal, communal or national self aggrandizement. Their conclusion is that, the greatest is God, that is the universal Self and all others are menial servants.

On a higher spiritual dimension, one is free from all kinds of manipulation, based on self interest. This dimension is attained by the foremost. "They are among them some who wrong their own souls, some who follow a middle course and some who are by Allah's leave, foremost in good deeds. That is the highest grace."[4] When one is free from any separate interest from God, all contradictions are cleared. One will find a universal harmony everywhere. He will see a divine, an arrangement for everyone. Handled by God's grace, one will work as an instrument without the necessity of any individual's selfish calculation.

---

4. Koran (35, 32)

By such kind of behavior, unlimited obstacles are removed. The right attitude is to reconcile all contradictions between believers through humility, service and dedication, not exploitation. With God, compassion and mercy are overflowing. When one enters that realm, one realizes that the only necessity is to work for God's interest. While the Koran calls this dimension the foremost, Vedic literatures describe it as *Upasanakanda*, devotional service, the process of restoring the soul's relationship with God by rendering unalloyed service to the Lord.

From the present position of egoistic boasting the soul has to progress through higher spiritual states. All mistaken thoughts have to be given up and replaced by accurate ones. The correct ones are given by God in the reveal scriptures. The Lord's teachings are open to everyone. Some people understand them properly, whereas others have great difficulty in grasping their real meaning. This inability is caused by different material coverings, which consists of various proportions of ignorance. How can one understand God, if the eyes and the senses are overly attracted by the charms of different material things? They have very little time to focus on God. Many prejudices cover one's attention. Busy with external transaction; the senses, the mind and intelligences are surcharged with other concerns. A lustful person looks for a beautiful body, a business man for a wealth, a householder for bigger residence, a farmer for a larger field of crops, a ruler for the means to dominate the subjects, a politician hankers for a great position and the religious hankers for heaven without God. Conducting false dealings, undesirable, alien elements cover the souls. The foes are all these false identifications. One is carried away by different waves of separate consciousness. Not fully connected with the Lord, having a separate interest, one is more or less separated from God.

The Lord is prepared to give his grace to everyone, but one must be capable of receiving it. He gives directions to everyone in terms of one's capabilities in taking guidance. When one refuses the Lord's guidance, one is entangled in the actions and reactions of his own action. In the Vedas, activities that neglect the direction of God are called *Vikarma*. Also, one may take guidance from the Lord with a motive to achieve a selfish personal goal. This dimension is called *Karmakanda*, motivated activities for personal material gain. Now, we shall explain how on the lower course of action *Karmakanda* and *Vikarmakanda* levels, devastating conflicts take place. Tragedies occur when patience, humility and compassion are lacking. For example, after the disappearance of Muhammad, overwhelming political tragedies in the form of civil wars between believers afflicted the Muslim community. In Vedic Literatures, due to *Daksas* enmity towards *Siva*, a terrible tragedy happened. Here, we are faced with two dimensions, the lower and a higher in a lower spiritual dimension, when one is entangled with the slightest difficulty, one takes shelter of intolerance. To alleviate any misfortune, extra ordinary plans are made to skillfully manipulate and exploit the environment. The atmosphere becomes surcharged with intolerance, envy, vengeance, quarrel, pride and false prestige. Egocentric designs result in a hostile atmosphere between believers. After Muhammad's death, due to selfishness, millions of believers killed each other, on the other hand, Joseph did not resist the plans of God. He never lamented for his misfortune. He did not waste his time in quarrelling, fighting or trying to harm his wicked brothers. He took help from God, exhibited patience and maintained faith in God. He concentrated all his thoughts on God rather than wasting his valuable energy in fighting with the antagonistic environment.

PATIENCE AND SANITY; THE PROMOTER OF PEACE AND HARMONY:

As is obvious from our discussion, all great wars or conflicts happened in the spiritual field causing a great damage to the environment and ecology. Due to lack of patience and sanity great wars were fought, nuclear weapons and chemical weapons were used in huge quantity, causing a great damage to ecosystem and environment. Hence, it is the patience and sanity of men, which are the basis, on which a sound environmental and ecosystem can flourish. In spite of so many spiritual conflicts, the great spiritual masters have shown abundant patience and sanity to protect social harmony, environment and ecosystem. We may now just give a paradigm case of such patience and sanity.

Joseph, the eleventh son of Jacob tolerated many tribulations with great patience. As a small child, his brothers out of jealousy threw him in a well. He was then sold as a slave and unjustly put in the prison for many years. Yet he did not take any revenge against those who harmed him, rather, he forgave them. The Koran praises the activities of Joseph. The story of Joseph has a very deep meaning. In it, there are many instructions for one, who is endowed with proper spiritual meaning. There is spiritual guidance and knowledge of how to achieve the mercy of Lord. The story of Joseph is full of meaning for the seekers of truth. Joseph said to his father, "Oh my father, I see eleven stars and the sun and the moon, I saw all of them prostrate themselves to me."[5]

Jacob said, my dear little son, do not relate this vision to your brothers, who may concoct a plot against you. The Lord will choose you. He will teach you the interpretation of visions. He will impart his wisdom upon you as He did with Abraham and Issac. "The Lord is full of knowledge and wisdom."[6] When Joseph attained his full manhood, God gave him power and knowledge. Jacob, the grandson of Abraham and Sara, and the son of Issac is the father of the twelve patriarchs or tribes of Israel; Jacob and ten sons from his first wife: Reuben, simeon, Levi, Judah, Dan, Napthali, Gad, Asher, Issachar and Zebulon. And two sons from the second wife: Joseph, and Benzamin. The ten sons of the first wife were envious of Joseph and Benzamin. They said "Our father has more love for Joseph and his brother Benzamin."[7] One of them replied that "In order to get our father's favour, let us kills Joseph and cast him away to some remote land."[8] Another son said, "Do not kill him, but rather, throw him down into the bottom of a well, so that a caravan of travelers may pick him up."[9] Then they went to their father, asking him "To allow Joseph to come, play with them in the forest. They will take good care of him as his sincere well-wishers."[10]

---

5. Ibid (12:4)
6. Ibid (12:6)
7. Ibid (12:8)
8. Ibid (12:9)
9. Ibid (12:10)
10. Ibid (12:11-12)

While taking him with them, they all agreed to throw him down into the bottom of a well. In the early part of the night, they came to their father, all weeping. They said to their father that, they went racing and left Joseph behind with the baggages. A wolf devoured. They stained his shirt with false blood. Jacob could understand that his ten sons were not speaking the truth. He said to them "For me, patience is the most befitting. It is the Lord alone whose help can be sought."[11]

Later, some travelers in a caravan, wanted to quench their thirst, found the young boy in the bottom of the well. After rescuing the child from the well, they sold him as a slave to a rich merchant in Egypt. While staying in his house, the wife of that wealthy merchant became very much attracted to Joseph. She tried to seduce him, closing the doors; she said, "Now come to me. Joseph said, your husband has made my sojourn agreeable. No good will come to those, who do wrong."[12] He would have desired her, but the Lord protected him. He would have desired her, but the Lord protected him. Joseph resisted the offers of his master's wife. The husband suddenly arrived at the door and the wife for rescue, accused Joseph of trying to rape her. Joseph claimed his innocence. A neighbor who saw everything bore witness to Joseph's virtue. The news spread out to the ladies of the city and when she heard of their malicious talk, she invited them for a banquet. While they were taking their meal with their knives in their hands, she called Joseph to come. When they saw him they were so amazed that they all cut their hands while looking at him. She said, "This is the man, for whom you did blame me. I tried to seduce him, but he declined. Now either he accepts or he will be cast into prison. Joseph preferred to be thrown in prison, rather than accepting her proposal. He said, Oh, my Lord, the prison is better for me than what I am invited for."[13] While in prison, two young men were with him. One of them said, I dreamt that I was pressing wine. The other prisoner said, I dreamt that I was carrying bread on my head and birds were eating. Please tell us, what their meanings are. Joseph said, I will surely

---

11. Ibid (12:18)
12. Ibid (12:22)
13. Ibid (12:33)

reveal to you the truth, of what will befall you. My Lord had taught me this knowledge. Of the two prisoners one will pour out wine for the king, the other will be hanged and the birds will eat his head. The first prisoner was saved. Joseph said to him, please request the king to pardon me. But the first prisoner forgot to approach the king and Joseph loitered in prison for several years more.

After Joseph had spent a few years in prison, the king of Egypt had two troubling dreams. He saw, seven fat cows devoured by seven lean ones; seven green years of corn and seven others dried up. The king asked his chiefs to interpret his dream. The first prisoner who was serving drinks to the king recalled Joseph and informed the king for the interpretation of dreams. The king called Joseph who fruitfully interpreted the king's dream: there will be seven years of abundance followed by seven years of famine in the kingdom, which became true later on. The king was so impressed that he made Joseph his viceroy. Later on, Joseph became the second in command of Egypt and the manager of Egypt's grain stores.

Later on, the hostile brothers of Joseph surrendered to him at the period of famine. Joseph excused them and gave them shelter. This prove that Joseph as a synthetic and broad mind in spiritual field, which is highly existential to keep the surrounding and circumstance in tune.

I have taken so much pains and time to narrate the above mentioned story to prove that in the field of spiritual dealings, abundant patience, synthetic mind, clear vision, bereft of all superstitions and extremes are highly essential to promote humanism resulting in promotion of ecosystem and environment.

While we speak of humanism we find different definitions and different narrations. The theory humanism is not also free from conflicts and debates. The two metaphysical models which torture humanistic thought are (i) Theism (ii) Atheism. Some humanists are theists and bringing humanistic thought to the extreme and creating some problems for the society, environment and ecosystem. Some other humanistic thinkers are atheists and bringing their theory to the extreme and creating some problems for the society, environment and ecosystem. Of these two schools we find several definitions and narrations, which are most confusing and devastating to promote social wellbeing, environment and ecology.

We can very simply say that, a man is neither a mind nor a body, nor an intoxicated theist, nor a reckless atheist. Man is something more than the above mentioned negative thoughts. Question then arises, what is a man? A man is a social animal, a rational animal, which has a sophisticated civilization, innovative progress, who lives on some value system for the promotion of all in the society. From the time immemorial, man has tried his best to promote the wellbeing of the society, although there is a continuous fighting in between the good and evil throughout the system of progress. The good will of man has gone through several tests and rectifications throughout the ages, still beholding forward to attain the utopia of perfects and fully fledged wellbeing of the society. Perfectness, although a utopia, is most essential to draw the attention of men towards goodness and wellbeing of the society. And till now, we are submerged in the race of wellbeing and goodness, because we are men. Our aim is to make our society rational, harmonious, synthetic and value oriented. Man is normal, rational, social, innovative and creative animal who has the propensity to protect environment and ecosystem intact. Unless and until this idea prevails everywhere in the society we cannot protect the society, the environment and the ecosystem.

Crores of rupees are spent, thousands of seminars are held thousands of projects are undertaken; can we be able to keep us free from eco-hazard? Certainly not. This is because of the fact that our inner sight has not been rectified. Our philosophical outlook has not been cleared up. Our cultural aptitudes conflict with each other. The day, all these conceptions be well settled, then only, we will be able to keep our society, environment and ecosystem intact.

****

CHAPTER-III

PHILOSOPHICAL THINKING OF MODERN MAN IN TERMS OF MECHANISTIC VIEW

PHILOSOPHICAL THINKING OF MODERN MAN IN TERMS OF MECHANISTIC VEIW

During industrial revolution in Europe, different types of industries were set up; things were produced in large scale which sought great markets for selling of their products. So, those were exported to different parts of the world. People learnt the use of those goods and imported more and more. As a result, they became richer and set up more industries. Some countries of other continents also set up different type of industries by following them. Human life became comfortable and easier by using different goods. Gradually they became dependent on machine. With the rise of industrial revolution, pollution rose simultaneously. It's bad impact could not be marked at the outset, as it was in the mild form.

On the other hand, as the things were produced in large scale by machines, it became cheaper. The industrially developed countries took the raw materials from the different parts of the world and manufacturing various items sent back to these countries by labelling higher rates. In the name of business, they started colonialisation in different parts of the world. British and French people were on the front line, though Dutch and Olandaj have tried to some extent, but could not succeed in their mission. In the principle of survival of the fittest, Britishers drove out other Europeans and ruled. If we look back to history, India was trapped by Britishers by driving out the French in the battle of Palacy. The Indian dominions, at that time were segregated and involved in quarrelling among each other, which gave the opportunity to conquer one after another. With the policy of divide and rule, they took away the independence of Indian sub-continent. The rules and regulations for administration were framed in British Parliament. As our sovereignty was lost, we could not rebel against those laws, which did not suit us.

A lion's share of resources from the Indian sub-continent was exploited by applying military power. Our glorious cultural heritage, agriculture, forestry, mines and minerals and oceanic products etc, everything was exploited and nothing was left from their greed. They took away the raw materials and sent back to us in the form of finished product. The Indian cottage industry faced a great loss and became weaker.

Some people felt dignified by using the foreign products and they were called aristocratic families. At that time, Kings and Jamindars were ruling the people by paying tax to the queen of England. They were the vicars of British Government. They needed the workers to serve as clerks, who will help in functioning of the government work. So, educational curriculum was framed accordingly to create clerks who will assist in administration. Gradually our mechanical way of thinking was sprouted with the adverse impact of foreign culture.

In this chapter, I shall focus on different points that how mechanistic thinking grew bypassing the philosophical thinking which were vitalistic in character due to amalgamation of other culture.

GURUKUL PATTERN OF EDUCATION IN INDIA:

During the period of long rule in India, Britishers imposed their materialistic and exploiting technique in every footings of Indian life that we are till now following those suicidal rules and regulations, even after six decades of independence. Our educational curriculum is not yet changed. As a result, most of the students are becoming unemployed after completion of education. But if we go back to our Gurukul pattern of education, we can see the pragmatic value of education. Pupils were practising each and every lesson with their Gurus and acquired the real knowledge about the subject. Pupils were greatly inspired by their Gurus on value- system and practical applicabilities of education. If we look into Mahabharat, the great epic of Hindu Religion; Aruni, Upamanyu and Arjuna were the burning examples of Gurukul system of learning.

But if we see the present system of education, such types of legendary personalities are rarely found. Student unrest, class room firing, rape, murder, etc. are found in present educational system. Even professors are shot or stabbed by the students. Sometimes the Indian students those who go to foreign countries for higher education, are killed by their fellow mates. But, if we think about Nalanda University of Magadh (at present Bihar), so many disciples from eastern Asia were coming for higher education and return to their native places, by praising the system, as a result more and more students from outside the country were coming. At that time more emphasis was given on quality of the education and interest of the students. A student who is well in literature may not be well in other science subjects. But in the present, one has to read all the subjects prescribed by the Board/University to obtain a certificate. Here students are reading mechanically against their interest. As a result, they fail or simply pass without any practical knowledge. Research and innovative works will not take place in the absence of talent of skill, which is now seen rarely among the students. If a student has no interest in a particular subject, how can he create something new in that subject? There is a saying that "Jack of all trades master of none". So he will be just a certificate holder having no real knowledge about it, as a result the number of educational unemployment will be more and they will be of no use for the society, rather a burden.

Though some sort of changes are found now-a-days in our academic system, with the rise of technical institutions, our guardians are not so much accustomed with it. On the other hand, the fees of such educational institutions are so high that, poor parents are not able to admit their sons/daughters in those institutions. Keeping in view of these points, Gurukul pattern of education is free from maintaining greed, superficial gain of certificates and mechanistic thinking.

Now-a-days, education has become an object of trade and commerce. The relation between teacher and student is measured through the fiscal capacity of the guardian. Love, affection, dedication, devotion etc. are on the wane. Therefore murderers, dacoits, terrorists are created after completion of their academic career. The political leaders, administrators or future descendants, who will be borne out of such type of educational system, will be of that type. We cannot expect mango from a Neem tree. So the environment of real education is destroyed. It is because of adverse impact of western civilization and education.

If somebody will say that western people are technically more developed than us, I cannot agree with him. Because, our forefathers were more developed than them. When we visit the Sun Temple of Konark and Lord Jagannath Temple of Puri, we feel proud about our predecessors. The sculptures on the stone are done so meticulously that, these are unique in the world of art and architecture. Even, it is beyond imagination that, how our ancestors could lift such big pieces of stones so high and how could they collect it? At present, we are unable to simply repair it with the same stone, rather patching with some other materials. The Taj Mahal of Agra is also another monument of Moghul period, which is within seven wonderful things of the world.

Again, in the field of mathematics, the contribution of Aryabhatta, Barahamihira and Leelabati are unforgettable. They have done outstanding works in the field of mathematics and astrology also, during the rules of Vikramaditya. Samanta Chandrasekhara was the great astronomer of Odisha, who could measure the distant mountains and study the movement of planets with the help of some bamboo sticks and pipes. These are the examples of our ancestors, who had done remarkable works in the field of science. They were educated in the prevalent system of Gurukul Ashram. The so called Gurus were *Rishies* or Sages. Our Indian sages were very much preservative and dedicated. Dedicated in the sense that, they have dedicated their lives for the upliftment and benefit of the human race. Their ideology was "*Vasudhaiba Kutumbakam*"; which means the whole world is one family. They thought that, in the development of a few people, world will not be benefited. We have to think about the mass. This holistic ideology made them worshipped by the whole. They were free from cruelty, hatredness, jealousness etc. and endowed with all heavenly qualities and always thinking about the development of human race. The disciples, those who were taught under them, made up with such ideology; for which environment of that time was in tune. At that time, character building was a main factor among the students. Therefore, mischievous activities were not found among the students, which is a major problem now-a-days among the students. As the students of that period were bestowed with all good qualities, they became the good citizens in their real lives. So families/societies were free from all sorts of corruption which are teasing us in every now and then.

It was seen that different *Rishies* were expert in different branches. A student after completion of a general course from one *Rishi* goes to another *Rishi*, who is expert in another subject. The student acquires knowledge about that from him. In this way, a student becomes expert and utilizes his knowledge for the benefit of the society, which keeps the society and environment free from corruption by bringing balance to it.

INDIAN SYSTEM OF MEDICAL SCIENCE:

In the field of medicine and surgery, our forefathers were not less than the western people. Charaka in his writing *'Charaka Sanhita'* proved the medicinal value of trees, found in the environment and said God has created everything for the well being of the mankind. We should not destroy anything. But we are cutting trees lavishly in the name of urbanization, industrialization and bringing natural calamities by causing various devastations. On the other hand, when we are clearing forests, different species of medicinal plants are destroyed. The indigenous people of forest area make their treatment with these plants. They do not take medicines.

In Indian tradition we have developed a kind of treatment, which is termed as *'Ayurveda'*. It is very much effective and sustainable having no side effect. In other words, it is eco-friendly. But with the development of modern science, another type of treatment, i.e.; allopathic treatment is developed which gives us quick remission, but it has various side effects. Again another type of treatment is there, i.e; homeopathy, which is sustainable and cost effective. Our forefathers had adopted that ayurvedic treatment for remission from diseases. The persons, those who were doing such type of treatment were called *"Vaidya"*. They were preparing the medicines from roots, fruits, leaves, barks and seeds etc. collecting from jungles. These types of medicines have no side effects. Even, if surgeries were made by *Sushruta* with the help of these medicines. This type of treatment was discovered by our Indian sages. But now we are adopting the allopathic treatment for immediate relief. As a result, we are facing various side effects and treatment continues. That happens, because we forget our own systems by saying orthodox. But in the name of modern treatment, we buy different types of other diseases, which are the result of mechanical thinking of modern man. We are saying that our system of treatment is good but in practice we are doing other ways. The most striking factor in this regard is the modern set up of medical science which induces the man to extract huge chemicals from the lap of nature creating thereby a tremendous loss to environment and ecology.

INTER MIXTURE OF RELIGION AND CULTURE:

After establishment of colonial base in various parts of our country, our local people were gradually attracted towards them; i.e., their education, culture, custom, habit, tradition, religion etc. when our people felt that those are flexible in nature, our young people were gradually attracted. Gradually some young people from rich families went to their countries for higher education, research etc. But out of them, some retuned and some did not. Those who stayed there, they set up families being tied up with marriage. Those who retuned they followed the foreign culture in our country. On the other hand during the period of British rule in India many English people stayed in our country. So our people were influenced by them either forcibly or persuasively or spontaneously due to intermixture of culture. At that time, some Christian Priests came to India for spreading of Christianity by alluring poor people. They selected economically backward areas especially tribal areas and started preaching Christianity. During that time, untouchability was in violent form among the Hindus. The lower class people were tortured in various ways. So they gladly accepted Christianity. When the conscious Hindu people knew it, they protested the Christian priests. Though there is law against forceful conversion or persuational conversion of religion, still the Christian priests did it by hiding them. As a result riot, massacre, burning of houses, loss of national property were found, which we have seen in the Kandhamal district of Odisha.

On the other hand, being attracted by our purity, serenity of Hindu religion and people from different religions all over the world made an organization, which is called *International Society of Krishna Consciousness* (ISKON). A well decorated temple is built in Mayapur, Nadia district of West Bengal, where Sri Chaitanya was born. People of various fields are gathered here spontaneously being influenced by the philosophy of A.C. Bhaktivedanta Swami. ISKON temples are built all over the world, where idols of Radha and Krishna are worshiped.

Again in the car festival of Lord Sri Jagannath at Puri, people of different religion, caste, creed, culture take part and dance by chanting the hymns of Lord Sri Jagannath. All are united forgetting their differences. Once upon a time a muslim, named Salabeg a great devotee of Lord Sri Jagannath had written so many bhajans and jananas and afterwards, he was known as bhakta Salbeg. It is a matter of great astonishment that, the calendar which is followed in Sri Jagannath Temple of Puri is prepared by a muslim, viz, Aminul Islam and it has been preparing on ancestral basis by his successors.

There are certain places in Odisha, where Durga Puja is observed by both Hindus and Muslims. So also Maharam is observed unitedly. Such type of intercultural tolerance brings friendship, fraternity among the communities and promotes ways for environmental protection and eco-system.

When different races like Moghul, Afgan, and Turk came to India, a cultural amalgamation took place and Indian culture became enriched. As a result the great monuments like Taj Mahal, Kutab Minar and Red Fort etc. were built and remain as the land mark in world history. Likewise, English people not only administered us with iron hands, they have also contributed so many things for betterment of the Indian nation. Railways, Posts, abolition of Sati Pratha, widow marriage, establishment of educational institutions were done by them, with the help of some Indian reformers for the well being of our society. Britishers, during their stay in India, observed different ceremonies and functions according to their own customs and conventions. Some aristocratic families of our country were also invited to their functions. They saw it and try to follow it in different occasions. Especially Kings and Jamindars were on that line. On the other hand, those students, who were going to European countries for higher education, they also imitated their tradition, food habit, dress etc. during long stay in their country. After returning to India, they followed their style. Other people also imitated them, as they were highly educated. So, gradually the foreign life style entered into our native life style and dominated also. The most important thing to be noted here is that when different religion and cultures go hand in hand in the society, there is progress of communities and protection of environment, but when religious ideologies and cultural whims clash with each other, the society is endangered resulting in a great erosion of environment & eco-system, which the history has seen from time to time.

THE ADVERSE IMPACT OF GLOBALIZATION ON OUR CULTURE:

Globalization has direct impact on outsourcing. Our young generations are running behind it, as if insects jump into fire. It has become more, just after globalization and liberalization in economy. In order to earn more, the young masses are going to foreign countries. In the name of earning, they are not only earning more money, but also various new diseases from their countries, like AIDS, Swine flu etc. which has no treatment or treatment are very much costly.

If we look into the history of these diseases, these are all outsourced from western countries. Illegal sexual relation is the main cause among other causes of AIDS. When it came to sex workers, it was spread over to different parts of country. Now-a-days it is found in third world countries and spreading rapidly also. Likewise, several precautions were taken by the government to prevent swine flu. The passengers, after getting off from the airplane are vividly checked up for detection of swine flu, but in vain. In spite of all sorts of defensive measures, the disease has spread to different parts of the country.

Here one thing to be noted here is that, we are spending crores of rupees to prevent AIDS. This is due to mechanistic thought. Had it been humanistic thought, we would have spent that money to create consciousness in order to create moral value in man to refrain illegal sex. There is a saying in Srimad Bhagabatam:-

> *Yatha tarurmula nisechanena*
> *Trupanti tatskandha bhujopasakha.*
> *Pranopaharascha yathendriyanam*
> *Tathibasarbarhana mechyutejya.*[1]

The meaning is that, if we will water at the root of the tree, it will grow and if we will take medicine in to the body, the limbs will free from all sorts of diseases. Likewise,

---

1. Srimad Bhagabatam, 4 : 31 : 18

if we make a man morally conscious, then he will be more humanistic and will keep himself aloof from disease like AIDS. If we will use crores of condoms, we cannot prevent AIDS.

DIFFERENT HEALTH HAZARDS DUE TO MECHANISTIC FOOD HABIT AND LIFE STYLE:

We have invented different types of machines to make our lives comfortable. As a result, we are gradually becoming ideal. Even if human beings are reluctant to walk. Therefore, different types of diseases, like high pressure, diabetes, acidity, cardiac problem etc. are commonly seen in the present society. Our forefathers, those who were working hard, were free from such diseases or rarely found.

Food habit and life style of the present time are responsible for above diseases. The green vegetables, which we are taking now, are hybrid in nature. They need more fertilizer and pesticides for yielding. So, poisonous particles remain in the food grain and enter into our body. Some immoral business men inject liquid sugar into unripe fruits and colour them by using chemicals and sell those in the market in high prices. Recently, while I was continuing my work, the issues regarding the production of B.T. Brinjals have created havoc in our country. An American company invented it, whose gene character is changed; it is less affected by insects and highly productive. But on the other hand, it is not good for health, i.e.; causing cancer. If it will be produced in our country, local varieties (350 types) will be seriously affected. Due to countrywide awareness, government could not allow it for production in our country.

It's result is best felt in case of production of high yielding paddy in our country. Different types of high yielding paddy varieties have driven out our native varieties. The native varieties were produced by applying organic manures, like cow dung and house hold garbage which are eco-friendly and do not create health hazards. But the mass use of fertilizers and pesticides create so many problems, like health hazards, food poisoning and bring remarkable pollution in environment and eco-system.

THE IMPACT OF AUDIO VISUAL AIDS ON YOUNG GENERATION:

The uses of audio visual systems are deteriorating the moral and environmental value of our young mass. Their minds are fickle and imitative. They also try to imitate the actions, styles of actors and actresses of cinema and T.V. The producers and directors of film and TV serial promote different lusty scenes for greater commercial purpose. Some female artists join their hands with them for their better advertisement and to earn more money. By following them, the young chaps of college or university level do accordingly in order to draw the attention of others. Some also try to get chance in film/T.V. They adorn themselves in the shade of artists. Different companies for the advertisement of their products, also take the help of actors and actresses, who shows their bodies in the name of advertisement to earn lucrative amount of money. Though different social organizations, female leaders oppose it, still they are doing without caring them. The actresses those who are taking part in these activities, they think that they are out of this society. Recently, it is seen that so many modern odia songs, bhajanas, jananas are written in dual meanings in the name of Lord Sri Jagannath, which have hurt several odia people. Therefore, it is decided that album songs will be sent to censor board before release. But it will not be so much effective. We have to change our mentality. If our mentality will not pure, then no censor board can control it. Rather, it will destroy our culture as well as environment.

In the recent time, mobile phone technology is developed in order to make our life easier. No doubt, it is a quick means of communication. If somebody is in distress, he can contact his near and dear, just at the moment. So, from positive point of view, it has so many good effects. But on the other hand, there are so many bad effects, which affect the society seriously. Theft, murder, kidnapping, sending of unpleasant messages and other sorts of offences are done by antisocial by using this technology. Even, it is creating problem for maintenance of law and order. If somebody questions, when such type of facility was not there in the society, were these crimes not occurred? The answer is 'yes'. But the ratio was comparatively less. Due to quick means of communication, the anti-socials are united quickly and run away being informed about police attack.

Some treacherous persons, by establishing good relation with the innocent girls, take up their photos of various postures and send it to other mobile phones or internet, where the girl is compelled to commit suicide. It is seen time and again. When the matter is disclosed, it is out of control. Cyber crimes are creating headache after the mass application of computer in government and non-government sectors. Before application of computer, everything was done manually. Now the data, entered in the computer are hacked by different agencies. If such types of occurrences happen in Defense and Foreign Affairs department, what will be the fate of our country? The foreign countries always strive to receive the information by alluring the dishonest officials to know the plans and programs of our country. But, if these works would be done manually, it may take more time or impossible to commit such type of crimes and the culprits would be caught easily.

Another type of theft is going on in banking system, i.e.; stealing money from ATMs. In most cases, the thieves are expert in using computer. However, the intension of using such type of facility is to provide prompt service to the persons during exigency. But stealing is done in such a sophisticated way that, it is not easy to catch them and they can take away in a very short span of time. Though thieves were seen in ancient times, but they could be caught and punished. But now-a-days it is not easy to catch them. Such types of crimes are seen among our young mass as they are expert in using this technology.

Therefore, the technical aspect of the audio visual system has created a great threat to environment and eco-system. For example, the establishment of mobile towers and setting up satellites has direct impact on environment. It is already detected that a mobile tower creates great health hazards to the neighbors and also affects the environment.

THE RECOGNITION OF HOMOSEXUALITY AND STRONG CONTROL OVER CHILD BIRTH ARE THREATS TO THE HUMAN RACE:
Few days back, homosexual relationship was recognized by different countries of the world. Where a man can marry a man or a woman can marry a woman. Does it not go against the law of nature? It was a time, when such type of work could not be thought of. It was treated as a kind of offence.

It will hamper the birth rate of the human race. In some countries, like Japan, China etc., the birth rate is so controlled that, after some years their countries will be filled up with old persons. Less numbers of younger would be there. So, governments of those countries are thinking about the implementations of development programmes. If the numbers of young mass will be less, then who will take care of the elders? Now the people of China are so accustomed with one child system that, in spite of government incentive, they are reluctant for the issue of second child. In some other countries like Russia, England, Germany etc., women are reluctant to be mother. They prefer to remain unmarried. Due to invention of various contraceptives, they are satisfying their biological urge without giving birth, If such type of tendency will grow, the family system will be collapsed, which will affect our environment and ecology.

THE ADVERSE IMPACT OF LIVE-IN-RELATIONSHIP JUDGEMENT OF THE SUPREME COURT OF INDIA:

Recently, honourable Supreme Court of India opined that, if two persons of opposite sex live together without marrying each other that does not go against the law. Because, Indian constitution has envisaged nowhere regarding its illegality and also said that, all immoral actions are not illegal. In Hindu mythology, Radha and Krishna were living together without being married. Such type of judgement was given by the Supreme Court comprising the bench of three judges, viz, the Chief Justice Mr. K.G. Balakrishnan, Justice Mr. Deepak Berma and Justice Mr. B.S. Chauhan in an appellate petition filed by Khusbu, a South Indian heroine. In 2005, she stated that, one can keep sexual relation before marriage. After her statement, twenty two criminal cases were filed by the different persons of the society, saying that it will have bad effect on the youths. She filed a case in the high court of Tamilnadu, in order to cancel the cases against her. But, the high court of Tamilnadu did not accept the case. So, she went to the Supreme Court and court ordered in her favour by saying that in Article 21 of Fundamental Rights of Indian Constitution, man has the right to live freely. When the lawyers of the opposite side argued that, it will have bad effects on the moral value of youths, judges told that, you prove, on which act, living together before marriage and enjoying sexual relation is illegal. Again, bench asked that, being influenced by the statement of Khusbu, which young lady has left her house.[2]

Here, we have to mark that, when constitution was framed and laws were made, such type of problem was not found in the society. On the other hand the love of Radha and Krishna is not worldly. It was in the mythology and interpreted by various interpreters from different point of view. If such type of system will prevail in the society, the society will be ruined and the court verdict will expedite the activity. Therefore, it is said in Odia that:-

"Thile thau pache guna hazara,

Charitra nathile sabu asara".[3]

The meaning is that, in spite of all good qualities, if man has lost character, then all his good qualities have no value. The fear of immoral action was so acute that, there was no need to frame rule for it. But now-a-days, due to deterioration of moral value in the society, such type of rules are required. Honorable Supreme Court has cited the love of Radha and Krishna, which was commented by different people of the country. If, we will take them as ideal and live without getting married, then the society will be polluted. Our culture, civilization, and environment everything will be lost. If an unmarried woman gives birth to child, shall the society take it easily? The answer is 'no', which is best known to everybody. She will not get social respect. Because our culture, our society will not accept such type of activity. Rather it is taken as pornography. In spite of court judgment, our society will not give them recognition.

---

2. The Odia daily News Paper- The Samaj, dt. 24.03.2010.0
3. 'Barna Bodha' of Madhusudan Rao.

HYPOCRISY IN THE NAME OF GOD:

If we look the system of observance of rites and rituals, social ceremonies or functions, at present, the sanctity, the serenity and holiness; everything is lacking. In the name of modernity, the idols of Gods and Goddesses are made in such a way that, we hurt by looking them. Their appearances are changed according to the sweet will of the organizers. We should not forget that Gods and Goddesses were imagined during vedic period. So their appearances were prescribed in the Veda. We cannot change their appearances according to our own will. In Lord Ganesh Puja, we see idols are made of different types of articles like mustard, Tila, Biri, leaf etc. in order to make it decorative, which should not be done. It should be made in clay or metals like gold, silver or brass etc. which is prescribed in Veda. The rites and rituals are not observed properly. Only album songs of different kinds are heard, which destroys the devotional feeling of devotees. Western people being influenced by the moral value of our society are attracted towards our country, leaving behind all the material pleasures of their country. Most of them have realized that, real peace is not in material pleasure of the outer world, but inside us. We have to recognize it through self-realization. What our Indian sages have known from Vedic period. In the philosophy of Veda, Upanisad; self or Atman have played an important role. Realization of self is the most important thing. Because, it is an indispensable part of Bramhan or Supreme Being. However our philosophical thinking, religious approach, ethical values have made us pioneer in the spiritual field worldwide.

But the excessive modern outlook in this regard has jeopardized the society, for which the environment and the eco-system is imbalanced. In the name of worship, different offences are committed which create so many problems in the society and go against our culture.

THE IMPACT OF INDUSTRIALISATION AND URBANISATION ON OUR LIFE STYLE:

Due to industrialisation and urbanisation, huge amount of water is required. The ground water is becoming scanty, because they almost use underground water. In coastal areas, where ground water level is high, tanks and rivers are more in numbers, in those districts of our state, we face drinking water problems during summer season, not to speak about other districts which are in the western region that is away from the sea. When we look to the news paper from February to June, till the monsoon reaches Odisha, such type of news are seen everywhere. People, domestic animals, wild animals suffer a lot, due to want of water. When tanks, ponds, streams, rivers are dried up, people depend on underground water, i.e; wells, tube wells etc. But when the underground water goes down, they are dried up. People go to far places, where water is available and stay on the line. Not to speak about other animals. Sometimes, wild animals enter into villages in search of water and attack people, as a result they die or people die. People rebel against the government for renovation or digging of new tube wells. Though government takes certain steps, but cannot solve the problem of the whole area. Now it has become a tradition that, such type of problem covers the news paper at the beginning of the summer season. In western region of our state, it is very much precarious. Here much importance is given to industries. Industries are more valuable than lives in the eyes of the government. They extract water in gigantic form for their use. Industrialization gives rise to urbanization, where much underground water is required for maintenance of life. But due to lack of water people suffer a lot.

Water, which is the corner-stone of life, society, environment and eco-system; is misused resulting in a great harm. The motive of explaining the above mentioned facts is to point out the fact that, when man is thinking and doing normally in the society, everything is OK; the environment and ecosystem is balanced. When we start thinking and doing in abnormal way, the society is jeopardized, the ecosystem and environment is imbalanced.

MIS-MANAGEMENT OF NATURAL RESOURCES BRINGS THREATS TO THE HUMAN RACE:

The western part of Odisha is filled with mines, minerals and jungles. But the inhabitants are chiefly indigenous people. They depend on jungle for their livelihood. They earn their living by collecting different type of fruits, berries, tubers, honey, wax, gum, wood, fire wood etc. from the jungles. When government decides to set up industries by exploiting the mining ores, jungles are to be cut off. When such types of activities continue; people of that area protest vehemently. Because, they loss their natural habitats. The industrialists, those who use the natural resources, do not care for anything, except their own interest. They exploit the mines and minerals by using advanced technology, which are non-renewable. We will lose these resources for ever. They extract much more beyond stipulation. Flora and fauna of that area are affected seriously.

This is due to excessive mechanistic outlook, which induced man to spoil the natural wealth and propels towards a great loss of environment.

HUMAN RESOURCE NEEDS TO BE DEVELOPED:

The per capita income of Odisha is very low and it is described as poor state of our country. The starving picture of Kalahandi district, selling of baby due to acute poverty has defamed us in front of others. But, if we take an overview regarding natural wealth of our state, it is full with everything which other states have not. The only thing is the lack of developed human resource. Those who are becoming educated are outing for better job. Governmental incentives are very much essential to control it. Where human resource is developed, the per capita income is more. Hong Kong is the best citing example of it. It is a great trading centre. People from different parts of the world come here for business purpose. So, per capita income becomes more. Now it is seen that, different multinational companies are coming to our state for installation of industries and signing MOU with the government of Odisha. But their progress is very slow due to public protest, red tapism etc. These should be overcome through discussion with the local people. The main intention should be benefit of the people and benefit of the state.

Here, the most important aspect to be remarked is that, we must try to develop the quality of human resource, which can protect the environment and eco-system while setting up new projects.

It is ongoing process in our state that, people are going to other states, other countries as 'Dadan Sramik' to earn their livelihood. The brokers, who take them for work by saying different incentives, cheat them and exploit them in the foreign places.

They work very hard, get less amount of food, fall into various diseases; but cannot complain due to fear of heavy punishment. Some of them slip away, return to the native place and narrate the precarious situation. Then district or state administration comes to rescue them. Some of them return to home with various life killing diseases like AIDS or so, which create problem for them as well as for their family. Though, government has taken certain steps to provide food for work for these poor people under different schemes, but not becoming successful due to illiteracy, evil intension of corrupt officials etc. So, strong, efficient, benevolent administrators should be selected for successful implementation of these programs and for the upliftment of the poor people. They will work whole heartedly for the benefit of the poor and illiterate people, by which our state our country will be benefitted.

THE EMERGENCE OF YELLOW JOURNALISM DEGRADES THE VALUE OF MEDIA:

Media plays an important role to create consciousness among the people through audio visual aids, news papers, magazines etc. Different government and non-government agencies advertise their plans and programmes with the help of film heroes/heroines, players of cricket, tennis etc. to draw the attention of public. Newspapers print different kinds of news and send to the readers. Electronic media also highlights different kinds of news. Here one thing we mark, the news readers are losing the faith on these as it was on the past, i.e., two to three decades before, because the question arises about the veracity of the news. Sometimes journalists are mutilating the truthfulness of the news by taking bribe or so, which is called yellow journalism. The impartial news increases the moral value of media. It is the fourth column of democracy. But now some news papers have leaning on different political parties. They publish the news showing favor to them. But such types of mischievous activities were not found in the past. Because, they are the index of our country. We can know day-to-day happenings of our country as well as other countries also.

If they will give wrong information, they will lose their dignity. Due to certain mischievous persons, people are getting wrong news' about the fact, which mislead the common mass. It is also found that, some journalists are bargaining with the corrupt officials, businessmen, politicians, criminals to publish news in their favor. As a result, people are remaining in dark about their misdeeds and they expand their network. Though different types of awards or prizes are given to secure truthfulness in the field of journalism, still they are not free from yellow journalism.

Again, journalists very often face attacks from anti-socials. They are killed, their cameras are broken, their family members are tortured etc. In those cases, they should be given both governmental and non-governmental supports for their protection. In spite of these, the news agencies should be free from all sorts of fear and allurements. Media is a strong channel through which the society, environment and ecosystem can be protected. The government and different news agencies have framed some norms for the publication of news. A government or news agency should not publish such news, which will hamper the social set up; the environment and eco-system. Journalists and news agencies must be very much cautious and constructive while publishing news in order to protect the community, environment and eco-system.

DUPLICACY IN LIFE SAVING MEDICINE:

Another thing, we all know that, medicines are life saving and very much essential for remission from diseases. So, these should be produced in utmost care. There are several checks and measures to test the standards of medicines. Drug inspectors are there in the department of health, who check the standard and reports to the government. Government Issue licenses to different companies for production of medicine. Those who violate the government rules, prescribed for production of medicines, their licenses should be cancelled. In spite of different checks and measures, some dishonest companies are producing duplicate medicines for greater profit. Such type of case was detected in kantabanjhi area, in Bolangir district of odisha recently.[4] The corruption was spread to the different states of our country. Here, a question comes to our mind. What kind of business is it? Beasts of jungles are rather more rational than such type of human being. He is more violent than wild animals.

---

4. Odia Daily News Paper 'The Dharitri', dt. 12.10.2010

To earn money easily, to be millionaire overnight, he had done such type of inhuman work. Is that the value of his education? When the matter was detected, case was filed; the drug inspector of that area was suspended and so on. But the culprit is always cunning. He knows better about the consequences. As the legal procedure of our country is a delayed process, he can get enough time to subdue it by putting pressure. After getting so much education, we are forgetting the human qualities. Only the person, who is expert in that field, can perform it smoothly. He was educated and developed in the society of our country to serve the human being. He is taught 'Service to mankind is service to God'. So our ancestors have conjoined the grace and curse of God for noble deeds and misdeeds, to keep the society and environment in tune.

There is a proverb in English that 'ill got ill spent'. If we will earn money in unfair way, it will also be spent in unfair way. It will have bad impact on children and other family members in different ways. There is saying in Sanskrit that:-

'Chhidre sya nartha bahuli bhabanti'

Which means one mistake or bad deed will give rise to several bad deeds. Bundles of problems will come to us. We will be suffocated by solving one after another. These are all due to greed of earning more money as well as the result of mechanistic attitude. Sometimes it paves the way for graveyard to the society by spoiling the environment and ecosystem.

MECHANICAL WAYS OF LEARNING FROM CHILDHOOD:

Now another type of cross thinking is seen in the field of mass education. From the very beginning of school education, guardians are admitting their off springs in English medium schools to make their children expert in English. The intention is that, their off springs will get job in multinational companies and draw shining amount of salary. But we are spoiling the infancy of a child. The children from the very birth imitate the parents and other family members in every respect, like use of language, respect to elders, rites and rituals of family, traditions, customs, culture everything. But when he/she goes to school, faces problems everywhere. He/she has to learn new language, new rhymes and new tales, i.e.; everything new. After some days of practice and utmost care of the teachers and guardians he/she becomes capable to learn everything and compete with fellow mates. But the question is why he/she will take so much pain from the mother's lap to learn everything foreign? It would be easier to learn mother tongue, rhymes and stories of our own culture. Because he/she hears these from grandparents and other family members. There are so many rhymes, stories in our culture. Our cultural heritage is the richest among others in the world. Our Vedas, Upanishads, Ramayana and Mahabharat etc are filled with so many tales and stories. These are so mingled with our flesh and blood that, even an illiterate country man can recite it, only by hearing from others. These are also filled with moral values. So a small child can remember it, because he/she always hears it in its periphery. But it will be troublesome on the part of a small child to learn and remember the tales, stories and rhymes from other cultures which are written in English. So, small children are trying hard to remember these. On the other hand, they are built according to the western culture. When these children are grown up, they bend towards the western culture, where family life and social life are totally different from us.

The fact is that, from the very beginning, we are educating our children in such an artificial manner, for which a new generation is created having no value system and with an outlook to gain more and more matter, i.e.; mechanistic outlook, resulting in a great threat to environment and ecology.

OUTSOURCING BRINGS DIFFERENT VITAL DISEASES:

The meaning of outsourcing is that to get the work done in other countries. It is mostly found in America. As man power is very much costly in America, the companies of that country outsource the man power mostly from India and China to get their work done in cheaper rate. The workers get more salary than their own country, so they are attracted towards it. The companies, those who give shining amount of salary, give so much work load that; the employees do not get leisure for relaxation. As a result, they suffer from various diseases, like high pressure, cardiac problem, diabetes etc. Once a professor of cardiology has said in a T.V. Program that, heart problem was seen among the young persons due to change of food habit, restless job etc. The young persons, those who are working in these companies, they get less time. So they prefer to take fast food; though it is time saving, but rich with oil and spices, which are not good for health. As they do not get sufficient time to prepare their food due to heavy work load, they are compelled to take it. In return, they suffer from various types of diseases. Now-a-days, everything is done with the help of computer. As they work sitting before the computer, they become prey to such type of diseases. Different researchers have said that, those who are spending more time working with computer, they are becoming old very soon. So in the name of earning more money, we are also earning different types of fatal diseases. They do not get sufficient time to spend with their families and lives of those persons are becoming mechanical.

This mechanical way of life has created such a society in which protection of environment and ecology becomes a day dream.

HYPOCRISY IN THE NAME OF INDIAN SAGE:

Now-a-days another type of cross cultural thinking is seen among some monks in Indian society that, though they are looking as traditional monks, but in fact, they are hypocrites. They exploit the common people in the name of God and maintain royal life in disguise. We find such type of news in different news papers in many times under the heading like, 'Baba caught red handed' etc. Different types of illegal actions are done by them. In the name of religion, they cheat the society in various ways, like smuggling, sex abuse, minting duplicate currency etc. In certain cases, they are able to get political patronage, for which it is difficult to detect them. Their circles are so strong that, even they can manage to escape to foreign countries. Compare them with the real monk of our country; whose duty is to worship God, think always for the betterment of the common people, help the distressed, eager to acquire real knowledge and lead a plain life. In our culture, when we see a monk on the road, we bow down our heads before him for blessings. Their place is in the heart of the people. Everybody respect them, as they are the well wishers of the society. But now-a-days, due to appearance of these hypocrites, the real monks are deprived of getting prestige, because we cannot differentiate them from the hypocrites. This is only due to the thinking that how we can acquire maximum material pleasure. But our ancestors like Budha, Mahavir, Nanak, Sri Chaitanya, Sankaracharya, all have preached that material pleasure is not the real pleasure, rather it is the cause of bondage. In order to attain salvation, we have to seek spiritual pleasure, which is Sat, Chit, Ananda.

The whole matter is to be noted here is that, due to insurgence of dishonesty among the so called monks of various religions of India, we find social disorder, riots, wars, which in turn has created a great threat to environment and ecosystem.

MALAFIED INTENTION OF VESTED INTEREST PEOPLE IN THE NAME OF SERVICE:

Today, a type of thought is sprouting among the learned people to set up NGO, in order to work for the development of the people by obtaining aids from the national and international level. Government also registers them and framed certain rules and regulations for their administration. They collect aids from different sources, to provide service during the time of natural calamities. They work against different social evils, take up several awareness programs for the benefit of the people. But some dishonest persons are there, who collect money and kinds for this purpose from different government, private sectors of national and international level and misutilise them. Only a small portion is spent for the public and larger portion is swallowed by them. They are frauds. They exploit the sentiment of the public through certain advertisement and word jugglery, which are hollow in nature. Some dishonest politicians, administrators and other official staff co-operate them in their work by taking share. It is seen that, some administrators of higher level have set up NGOs where, their wives or relatives work as head, collect aid from different agencies in the name of social service. They spend a little amount for the purpose and swallow the rest amount. If we compare the persons of two different ideologies, that our forefathers had built Dharmasalas or Beat houses for the benefit of the public out of their own purse, but the later are looting the public money in the name of social service to be rich over night. The formers had pious thinking, later have ugly vested interest thought. Both are the son of the same soil. The difference is that, the formers were in some years back, the later are acting at present. The later are highly qualified than the formers, but lacking moral value.

The degradation of moral value is found everywhere, especially in the field of NGOs. They should have been prepared to protect environment, but on the contrary, due to mechanistic outlook, they exploit the environment for monetary gain.

CORRUPTION IN EXECUTIVE IS IN VIOLENT FORM:

Our state government and central government have separate executives. But they work, keeping co-ordination with each other. All plans and programs are chalked out by the top executives, i.e.; the secretaries, directors along with concerned ministers and then implemented. They get monthly salary to render service to the public and work for the country. But in reality, some of them do not discharge their duties properly. They take bribe from the public. People also give them to get their work done. It has become a regular practice in our society now. Though there are various government agencies to check and control the corruption, such as; police, CID, CBI, RAW and several others, still the corruption is increasing. If we look into the history of developed countries, they are developed due to their honesty, hard-working, truthfulness etc. Such type of immoral activity was originated out of mechanical thinking. By forgetting the human qualities, we have become money oriented and to get maximum material pleasure is our motto.

The mechanical thinking of ourselves has made us blind. So many officers, both at upper and lower level are caught in the net of various criminal investigation department, media through sting operation etc and lose their service. Not only service; their prestige, dignity, self respect, everything they lose. But in certain cases it is found that, those who are involved in controlling corruption, they are not free from corruption. They also take bribe in huge amount and slip of the guilty person. As a result, corruption is increasing day by day. But in certain cases, public protest against the corrupt official and take revenge on him.

A government official is the son/daughter of our country. Our country has made him/her what he/she is. Again, he/she is paid from the government treasury, to serve the nation. In no way, he/she is entitled to take the bribe. It is a criminal offence. The person, who is doing such type of work, is well aware about it. Recently, a case was detected in our neighbouring country Pakistan. A female IFS officer Madhuri Gupta (50 years age) was appointed in Pakistan by the Department of External Affairs of India. In course of time, she was allured by government of Pakistan and sent secret reports of our country to them.[5] The whole country was stumbled when they heard the news. When I read it from the news paper I could not believe it. Being a top ranking, senior and higher salaried person, if she will be involved in such type of work, then what to talk about the others. She kicked her motherland and forgot everything being allured by material pleasure.

Here a question arises, after acquiring how much of success a person will be satisfied? She had got everything; power, prestige, shining amount of salary, foreign tours, what a person desires. In spite of everything, she was dissatisfied. The want of money and worldly pleasure have made her mad.

Our ancestors like Budha, Mahavir, Sankaracharya have advised us to remain aloof from greed. In Mahabharat, Duryodhan lost everything without giving five villages to his cousin brothers. It is seen everywhere that, good and moral actions reap good result, bad and immoral actions reap bad results. Human beings are doing corruption knowing everything. He/she knows that, the day will come, when corruption will be detected and bring punishment also. Still people do corruptions without thinking about the consequences. It is only due to mechanical thinking and it has made human being blind and bereft of all good qualities. To earn more money and to get maximum material pleasure is the chief aim at present. Money has made them shameless, regardless and valueless creature.

The effect of this money oriented administration finds its way on the direct looting of forests, mines, water for the fulfillment of personal greed, which creates a great hindrance in the promotion of environment and ecology.

---

5. Odia News Paper 'The Dahritri' dt.30.04.2010

THE PROPERTIES OF GODS AND GODDESSES ARE NOT FREE FROM THE GREED OF MAN:

Another kind of mechanical thinking, which we find among the vested interest people, i.e; to take away the properties of different temples, abbies or other philanthropic associations in order to be rich. Our ancestors; kings, Jamindars have set up different temples and donated landed properties for their maintenance. Rites and rituals are observed out of the products of these landed properties. In this way, Lord Jagannath has enough properties, which were obtained as gift from kings, Jamindars and different persons in course of time. These are scattered all over the country. People of that locality cultivate those lands and pay some portion of it to Lord Jagannath. But, as they have been cultivating those lands on ancestral basis, they are not paying the share, rather trying to enlist these lands in their names. This is found not in the case of the properties of Lord Jagannath, but in case of other Gods and Goddesses also. When it was brought into the knowledge of the government, a department was formed in the name of 'Endowment Department', whose duty is to keep watchful eye on the properties of gods and Goddesses by forming trustees for their maintenance.

But, when the multinational companies from various parts of the world came to set up industries or universities in our state, government is supplying them such type of lands. When Vedanta Company was assured to supply such type of land along with other government and private land to establish a world class university near Bhubaneswar, people protested against it vehemently, for which the proposal was postponed. Before touching such type of properties, we should think that, these were donated by our ancestors for their maintenance, who are we to take decision on it? Those who have donated, their soul cannot remain in peace. Again, those persons, who are taking away such type of properties, are earning sins for their families.

Another type of behavior that is seen among the priest in various temples that, they exploit the pilgrims, in the name of God. They take advantage of religious thoughts of people and exploit them in various ways. If we visit Lord Tirupati Temple, different types of rates are fixed to visit Lord. A rich man can visit Lord immediately by paying a higher amount of fee; where as a poor man will stand in a queue for long period to visit

Lord. Such type of difference is fixed by the managing board in order to earn more money. They have made God, the matter of business. Religious thinking is replaced by material thinking. Such type of cross thinking is very much shocking for religious persons. The religious environment is gradually changed and polluted also. People, who go to religious places, have pious thinking and want mental peace through dedication. Those who have committed any sort of crime and realized it, they try to transform themselves by dedicating to God and vow not to commit crime further.

Here a man is changed by heart, and society is saved from his oppression. Therefore, religion is that which keeps up the man and thereby society also. Religion is beneficial in every aspect of human life in order to make the society orderly. But now-a-days, we are going away from religion in the name of advancement; we want proof for the existence of God. Some argue that, if we cannot see Him, there is no God etc. Such type of thinking is making man more materialistic. He cannot get shelter, when loses faith on Him and become disappointed at the time of disaster. If we would have reliance or faith on any supernatural being, then we would not be disappointed. God may be necessary not for His existence, but for the protection and systematic progress of society.

GREAT DIFFERENCE IN SAYING AND DOING WHILE OBSERVING SIGNIFICANT DAYS:

To-day a new type of culture is found for observation of various days, in order to create consciousness among the people for the preservation, safeguard and remembrance of significant persons or incidents. Previously, we were observing birth and death anniversaries of Gods and Goddesses, important persons, Independence Day, Republic day etc. People were observing from the core of their heart. But now the number has been increased, like Earth Day, Health Day, Environment Day, Food Day, Father's Day, Mother's day, Teacher's Day, Children's Day etc. No doubt, this is a good symbol; we should create consciousness among ourselves by observing these days. In those days, the speaker narrates various reasons for observance of these days and gives good suggestions etc. For example, a speaker while delivering a speech on World Environment Day gives so many suggestions about how to save the environment from pollution, for the very existence of the world. But if we look into the life style of that person, we can mark several opposite things, which are not at all good for the environment. We say so much in the Women's Day; for their equal status with men, safety, representation in several elected bodies, reservation facility in government and non-government sectors etc. But in reality, there is every dominance of men. In spite of several laws to suppress the women harassment, they are not free from it. The speaker, who is speaking so many things regarding child labor problem, might have a maid servant at his house to do the house hold work. So many children are struggling in hotels, restaurants and motor garages etc all over the world in order to earn bread and butter for their family. Even female children are engaged as sex workers in different countries. We enjoy their services, as if we do not know anything. Though, different organizations at national and international level are working for them. But it is not sufficient to control it. We also observe World Health Day. But pollute the food and fodder by using inorganic manure and pesticides in the crops for increase of production. Some dishonest businessmen apply different types of chemicals in the food stuff for preservation up to long period as well as to get more profit, which create health hazards.

So also World Earth Day is observed to take different precautionary measures for safety of the Earth. One of them is economic consumption of water. Because, the underground water is decreasing alarmingly. Even in the towns of Coastal Odisha, we are facing the scarcity of water, not to speak the western areas of the state. Due to exploration of mines and minerals, establishment of industries, the temperature is increasing gradually and varying from 45^0 to 50^0 Celsius during summer. The ground water level decreases, the lakes and ponds become dry during summer. The lives of those areas suffer severely due to want of water. When we open the news paper, we see the picture of long line waiting for a bucket of water, the death of wild animals due to want of water etc. When life is in such a critical situation, government gives permission to industries to take river and underground water for their use. Though they are permitted to a certain limit, but they extract much more according to their greed and the version, economic consumption of water is of no value. They also pollute the river water by leaving industrial affluent into it.

In ancient time; kings, emperors and rich persons have dug wells; tanks etc to provide water to the lives and thereby earned wellbeing for them as well as their families. These ponds and tanks also help to increase the level of ground water. But now, we are filling the ponds and tanks, for construction of houses or agricultural lands etc.

From the above discussion it reveals that, the mechanical thinking of men has made us hypocrites, we are saying one thing but doing another thing. We are preaching for the protection of environment, but in practice, we are doing so as to destroy the environment and ecosystem. If this will be our mentality, the observance of different days will remain in discussion having no impact at all.

IN THE NAME OF HELP TO THE DISTRESSED, WE LOOT THEM:

It is a fact that, when somebody is in distressed condition, the fellow beings come to rescue him. When an elephant is in danger, the leader of the team comes to rescue him. Likewise, the leader of the monkey group keep watchful eye on other monkeys coming under the same group. Therefore, they live in groups. Human beings know it very well. Family, society, tribe etc were formed for the safety and security of the human race. We enjoy the social functions collectively, so also help the distressed by forgetting all the anomalies.

But now-a-days, we see certain deviations in it. People from various corners of the world send the relief materials to distribute among the distressed persons, who are affected by natural calamities. But the persons, who are engaged in distributing the materials among them, take away something out of it. During the period of 'Mahabatya' in the coastal district of Odisha, it was found in violent form. When distressed people were crying for and dying without food, shelter, medicine etc, the corrupt persons were looting the relief materials to become rich overnight. In Kedarnath disaster, even the sages were involved in looting the money and ornaments from the dead bodies which were recovered by soldiers involved in rescue operation.[6]

They have thrown down all the human qualities and have become more dangerous than wild animals of the forest. Their education, conscience, fellow feeling, sympathy are on the wane. Material pleasures have made them blind.

In this respect, another thing is that our government is receiving crores of rupees from the World Bank and foreign countries for checking and preventing natural calamities. It is a matter of regret that, a large share of the fund is going to the pockets of corrupted officials, for which, we fail to avoid natural calamities, thereby endangering the environment and ecology.

INCREASE OF ACCIDENT AND LOSS OF LIVES DUE TO INDUSTRIALISATION:

At present, government of Odisha is taking steps to set up industries, for the economic benefit of our state and saying that our people will get more engagement thereby earn more money. But simultaneously, the rate of accident is increasing day by day. A report says, during the last ten years, six hundred people have died, due to accident in industries. The management of the industries is not giving importance on the safety and security. They should arrange workshops, for the creation of awareness among the workers about their safety inside the factory. Our people are acquainted with agricultural practice, but new in industrial field, so they are ignorant about the accidents inside the factory. A statistical data of Odisha says that, the death rate inside industries has grown from 2004. It was 39 in 2004 and remaining within 70 to 80 from 2006. In 2008, 81 people died in accident inside the factory. It was 39 in 2004 and remaining within 70 to 80 from 2006. In 2008, 81 people died in accident inside the factory. In 2009, it rose to 122. Most of the workers are dying in iron and steel industry. These accidents are artificial and could be avoided if certain precautions would be taken. In spite of several warnings given by the government to maintain safety inside the factory, the company authorities are not becoming so much cautious regarding the safety of the workers.[7]

---

6. Odia news paper, 'The Samaj', dt 26.06.2013.
7. Odia daily news paper, 'The Samaj', dt 17.05.2010

Here, we have to remark that industries are set up for the betterment of the people of our state, not for the profit of the companies only. If people will die like anything inside the industry, then what is the necessity of these industries? So, necessary steps should be taken against those companies, where accidents occur, for which accidents would be diminished and lives would be saved. While I was writing this thesis, the judgement of Bhopal Gas tragedy was declared, in the district judge court of Bhopal, after 26 years of occurrence, which shocked each and every Indian. It is just a mockery to the human race. In the midnight of 2nd December, 1984, when the whole city was in the sound sleep, the worst historical industrial accident took place in the Union Carbide Industry, when water entered into the gas tank. The safety valve was opened and the most poisonous ammonia gas was leaked out. Thousands of people died, lakhs were injured. Death rate was increased and estimated that 35 thousands of people died and more than 5 lakhs of people were seriously affected. Several people became blind and lame. The children, who took birth after the accident, became disabled.

At that time, the chief of the industry, Mr Warren Anderson, an American industrialist, went away to his country with the patronage of top political leaders of our country. Government of India claimed $ 330 crores, but got only 14.2%. Those who survived from tragedy got less than fifteen thousand rupees only. In the charge sheet of CBI, 12 persons were accused. Out of that 8 persons were Indians and 4 persons were foreigners. But Mr Anderson was not accused. He was declared as absconded. The case was filed with the article 304(a), like road accident. Nineteen judges have judged the case for twenty four years and 178 witnesses were examined.

The government of India finalized the case with only $ 47 crore with the American government, which has not yet reached to the distressed. It is a matter of great sorrow that, those eight persons are given only two years of imprisonment and one lakh rupees as financial punishment, Mr Anderson's name was not in the judgment. After the declaration of judgment, these persons were allowed bail within a short span of time, which is like putting lime in the wounds. When the public knew about the judgment, they protested vehemently. The government of India took it seriously, a cabinet body was formed, who suggested to give Rs 1500 crores as help to the distressed.

Apart from it, the corrupt political leaders, who were involved in manipulation of the case, should be brought to the fore front, for which, the people will able to know them. The role of the judiciary was not clear, it was very much susceptive. Though there is agreement of restoration between India and America, but America is not restoring Mr Anderson, who is the main culprit of the accident. Now it is the time, to think about the consequences and learn lessons from it.Before signing a MOU regarding establishment of industry, the first and foremost duty should be benefit of the public. In the above case, the public suffered too much, whereas the company earned $ 950 crores as profit in that particular year. America, who is shouting for safe guard of human rights, is hiding the culprit. Therefore, the companies those who are adopting modern technologies, taking safety measures, aware about the environment and conscious about the interest of the local public, only they should be given preference for the establishment of industry. But those companies, who are black listed, adopting old and abandoned technology, they should be avoided, though they are agree to give maximum profit. Government should take clean and fair steps, to judge the companies before signing agreement with them.[8]

Industrialization pushed to its maxim, ignoring humanity and human welfare, has several bad effects which pollutes the environment and eco-system in large scale. A philosophical and cultural atmosphere should be developed to make industrialization humanistic to safeguard the environment and ecology.

8- Odia News Paper, 'The Samaj', dt 15.06.2010

THE HORRORS OF NUCLEAR ENERGY:

Now, there are so many nuclear weapons in the different parts of the world, which can destroy the world, so many times. But when nuclear technology was invented, it was meant for the production of energy, which will be used for the betterment of human race. But certain despotic statesmen used it for production of nuclear bombs, in order to use against enemy country, i.e.; America threw atom bombs on Japan during Second World War and caused a great devastation. Gradually the technology was handed over to the third world countries. In the third world countries; government is not strong and stable, internal disturbance is a regular work, the terrorists always kill innocent persons as well as government officials etc. So who can say that, these nuclear bombs prepared by the country will not go to the hands of terrorists? If those will go to their hands, the consequence cannot be thought of. World cannot forget the horrible attack of 26th September 2004 on World Trade Centre and Pentagon and 26th November 2008 Mumbai attack by the terrorists. Australia and Japan have jointly predicted that, if there will be nuclear war between India and Pakistan, then the weather will be changed and it will have great effect on the agriculture of the world. The stratosphere will be severely affected by the great devastations caused by heavy bombardment on cities and other important places of both the countries. This has been declared by eighty scientists after research under the topic that, "The impact of nuclear war on the weather". The poisonous smoke will cover the earth for decades and sun rays cannot reach the earth, for which there will be nuclear winter. Many species of both flora and fauna will be extinct, one billion people will die due to want of food, people will quarrel among themselves for food and suffer from different diseases, which is predicted by an American medical expert Ira Helfand.[9]

Before the invention of nuclear weapons, the war was fought with traditional weapons. Some soldiers from both parties were killed, some injured, and properties of both the countries were lost. However, the destruction is limited and confined within

---

9. English daily news paper 'The Telegraph', dt 07.06.2010.

two countries. It's effect is not long lasting and the war devastated areas can be reconstructed within a limited period.

But the devastation caused by the nuclear war is not repairable. It can ring the death knell of human race. Now most of the countries possess the nuclear weapons. If one country will throw, definitely the other country will throw. So both the countries will be ruined. Some countries like America and Russia have nuclear bombs, which can destroy the world so many times. So the mechanical thinking in the field of war to finish the enemy country by switching buttons, will invite so many dangerous effects to its own.

So America and other developed countries are trying to neutralize the nuclear bombs. But the nuclear technology is not confined within the developed countries; it has gone to the hands of countries like Iran, Pakistan etc, where there is every possibility of misuse. While I was preparing my thesis, China made an agreement with Pakistan to set up nuclear power plant for the production of nuclear energy, in spite of grave concern expressed by India and America. Because who can say that, Pakistan will not prepare nuclear bombs. It is seen that, Pakistan obtained the technology for production of nuclear bombs from China secretly in order to threat India. On the other hand, it can be said that, Pakistan is the motherland of terrorism. All the great terrorist attacks were chalked out in Pakistan. According to the satellite picture, so many terrorist groups are working there. The government of Pakistan is helping them through ISI agency to create mischief in India. As their people are illiterate and poor, they take advantage of it. The young mass joins in their group in order to earn something. They join without knowing about the consequences.

According to the statement made by Kasab, the only living terrorist of Mumbai attack, he was educated up to class IV. Before joining in the terrorist group, he was stealing in different places. This is the scene of villages in Pakistan. In spite of deep economic instability, the government is spending for arms and ammunitions, war craft etc. Even from the foreign aids, they get for different developmental work, purchase weapons. So there is always the fear of misuse, which will create great threat for the mankind. America and other developed countries are expressing deep concerns about it. Therefore, the invention of nuclear bomb is not for the protection, but for the destruction of human race. Implementation of nuclear energy in production of weapons is the sheer expression of supremacy and arrogance of statesmen to become superpower and to have control over other countries.

Not only in the field of nuclear bomb, but also, in the field of nuclear set up, we have seen the dangerous effects, when the reactor is mismanaged; there is every possibility that the rector may radiate huge quantity of Gama rays, so that a portion of earth may be smashed. Nuclear invention, when used for the good for the humanity is highly beneficial, when misused, is highly dangerous paving a way for genocide.

The most important thing in this regard is to generate a philosophical and cultural aptitude in the society, for the betterment of humanity and humanistic value. America, who is now the pioneer of world-class is saying of philosophical and cultural value to protect the environment, but its thinking and acting goes totally against this culture. The thinking and acting fashion of America is totally mechanistic, which is hindrance for the promotion of environment and ecology. The saying fashion of America looks like vitalistic and value oriented, which has hypo critic ingredients. This cross-cultural aptitude, perhaps, is the chief reason for which the world is foreseeing a great environmental and eco-crisis.

THE GENEROSITY OF HINDUISM ATTRACTS ALL OTHERS TOWARDS IT:

After attainment of all sorts of material pleasures, Americans are leaning towards Hinduism, which says about the realization of self, the spiritual way of life etc. Individual self is the part and partle of universal self, which is God. They are gradually leaning towards the Indian belief that, as all the rivers flow towards ocean, so also all the religious beliefs tend towards God. They do not want to belief the version of the Jesus, as stated in the Bible, that "I am the truth, I am the way to reach the father". They believe the existence of soul, the relation between soul and body, and the immortality of soul, rebirth, which Indians believe since long and called Hinduism.[10]

Though 76 % of people are Christian in America, still 65 % of them faith on Hindu trend. Among three American Christians, third one believes in burning the dead body, he believes in rebirth. The place of realization is becoming stronger more than religion. To go to Church, Mosque and Temple; reading the Bible, Koran and Gita are called religion. But meditation is called realization. Religion makes us pure and true, paves the way for realization of God. Some say that, this realization of God is nothing but realization of self. When we know about our own selves, it is also possible to know other selves. The whole universe is created by the supreme creator. We forget the distinction with other selves and think one with them and identify with them. The nature is within us and we are within nature.

From this observation, it is clear that, though all Americans are not Hindus, but they are leaning towards Hinduism. Because such type of ideology is not Christian ideology, rather Hindu ideology. Jesus said that, His path is the only path to know God, but Hinduism said that there are several paths; Jesus's path is one among them. In Hinduism, emphasis is given on 'Yoga', which says about the realization of self. Our mind is fickle and fluctuating; we can control it through 'Yoga'. Therefore, it is called "Chitta-Vriti-Nirodha". Such state of mind helps us to think that, we are one among others, both living and non-living. Because, we believe that God is the creator of everything, the supreme creator.

Here we mark that, America, being a country of materialistic outlook is changing its attitude and leaning towards Indian thinking envisaged by Vedic sages. But, we being the inheritors of such thinking are turning our minds towards them and following materialistic path. As a result, we are facing various types of social hazards every now and then. Our ancestor Swami Vivekananda was praised in different western countries by preaching Hinduism. The generosity of Hinduism made him great. Its ideology is very much liberal and respects all other religious beliefs.

---

10. Odia daily news paper," The Samaj", dt 15.06.2010

Now-a-days, some people are criticizing 'Yoga'. The reason is that, if 'yoga' will spread, then selling of their items will be restricted. According to materialists, the world as a whole is a market. As America is a materialist country, the ideology of world as a market is more. Again, people of America are not born Americans, but Americans by their living. Such type of thinking is good symbol for us. In spite of several differences, we are Hindus. So Hindutwa is not confined within humanity. It has spread to all living and non-living beings. We should be proud of it. It was collected from an American magazine 'News Week' under the heading "We are all Hindu now" by Arta Mishra and published with his opinion in an Odia news paper 'The Samaj' on 15th June 2010.

What Americans are feeling today, our ancestors have felt it much earlier, i.e.; Vedic or Pre-Vedic ages. The way of life style was formulated by them. They have touched every sphere of life and formulated certain rules for the maintenance of happy and prosperous life. They did not deny the necessity of matter or things. As we are possessing material body, the necessity of matter is inevitable. But it cannot give us real pleasure. It can only be obtained, only when we lead spiritual life. That is called Hinduism. Our forefathers mixed religion with our life style, to have control over the society. People were living peacefully. But the impact of western culture on the Indian society has disturbed our way of life.

This cross-cultural system has sowed the seeds of materialistic thinking. Our young mass are blindly following their paths, imitate them in every footings of life by hating our tradition and culture and saying that, our system is orthodox and outdated etc. But this article "We are all Hindu now" will be able to open their eyes, who hated Hinduism. Because, those who have become model of our young mass, they are following our culture. In Hinduism, no one can claim that he is the founder, as Mohammad in Muslim and Jesus in Christianity etc. Hinduism is the way of life. So being an old religion, it has farsightedness, which is liked by people of different time. During the old time, superstitions and blind beliefs were more among people. So these were mixed with religion for wide acceptance and more observance.

In Hinduism, nature is worshipped in different forms; like snake, turtle, fish, tree etc. which means all are treated equally in the same line as human being. If they will be sound, the environment and ecology will remain balanced and earth will be free from various forms of hazards. In this regard there is Vedic saying in the form of *santipatha* that:

> "Om Dyo shantih, Prithivi shantih , Apa shantih, Ausadhaya shantih,
> Vanaspataya shantih, Viswadeva shantih, Brahma shantih,
> Sarvam shantih, Shanti re va shantih, Sa Ma shantihredhi"[11]

This vouchsafes the Indian tradition of protecting and prompting ecology and environment.

INDIAN RELIGION COINCIDES WITH SCIENCE:

In great Hindu epics, like the Ramayana and Mahabharat, we see the use of different kinds of things, which the modern man is using now with the advancement of science and technology. In Ramayana, it is described that, Ravan had stolen away Sita with the help of Puspaka Vimana; which can be compared with helicopter at present, though some say that, it was a myth only, it had no physical existence. We may argue that, if it had no existence, how could it come into the mind of 'Maharshi Balmiki'. So, it is ontological that, it had existence. Sri Ramachandra rescued Sita with the help of monkeys, bears etc. Now-a-days, we see that, America is preparing monkeys with proper training to fight in the battle field or in radioactive areas. In Afganistan, Talibans are giving training to monkeys to involve in terrorists attacks. So, we cannot say that, epics are full of myths; they have no historical base etc.

The point is that, scientific advancement was there in the past and now also we see scientific advancement, the difference is that modern man has made the inventions of the most harmful weapons, i.e.; nuclear weapons open, for which, the world is running towards destruction.

---

11. Sukla Yayurveda.

A COMPARISON OF PAST CULTURE WITH PRESENT CULTURE:

Under this heading, I have cited different instances, where our past culture is compared with present culture. Our past culture is beneficial to the society, where as different kinds of hazards are seen in with present culture which endangers our environment.

During British rule in India, they employed their people in our country for administration. They recruited our people for clerical work, to help them in the administration. Kings and Jamindars of our country were working as their agent and collecting revenue for them. Where some British people were appointed, they had set up clubs and other recreational centers for their merry making. The Indians, those who became intimate with them, they also went to clubs with them. Gradually their culture devoured our culture.

At that time, there were 'Bhagabat Tungis' in every village. People go there at the evening and hear Bhagabat, the great epic of Hindu. It was in poetic form and full with deep morality. People learnt it by hearing. It had also a great influence on human being as well as society. Our people were quoting it's rhymes during conversation and obeying it's inscriptions. Even an illiterate person could remember it by hearing frequently. Sometimes, the farmers were singing it's rhymes, while working in the field in order to lessen their tiredness. From this, it can be understood that, how deeply it was rooted in the minds of Indians. Therefore, we did not see corruption, that we are facing in our daily life now. But, when the club culture rushed into our society, the 'Bhagabat Tungis' are gone and we are mourning for the past.

If somebody will argue that, it is the human tendency to praise the past. He is partially right. Though we have developed so much in science and technology, we have invented so many things to make our life easier, we have controlled our food problems and epidemic etc., still we are feeling the want of something, i.e.; morality, humanity etc., which bound our ancestors so tightly that, in spite of acute poverty, they were maintaining peaceful lives. But, now we are crying for peace.

Previously, joint family system was there in our country. In spite of different troubles, joint family system was preferred. One came to rescue of other, when someone was in distress. The old and senior persons in the family did not feel helpless. They were given proper service. But now, we see old age home, where old men are living due to lack of proper care from their family side. Can old age home be equal with family? Can an old man get real pleasure living in it? The answer is definitely 'no'. Because he is compelled to live in a new place, living behind the most favorable house, which he had built in his own hand and carries the experiences of different stages of life; infant, childhood, youth, adult etc. He would only express his sigh and count days of the rest part of his life.

It is not that those who have gone abroad for service, their parents are facing such type of problem, but some sons, after getting married, prefer to stay outside by neglecting their parents. In those cases, the sufferings of old parents are intolerable and inexpressible. In remote village areas, it is very much difficult for them, because food stuff and medicine are not easily available. In certain cases, they have no money to meet their necessities. But, now-a-days, governments are giving old age pension to meet their necessities. In reality, that is a drop in the ocean. If this will be fate of our elders, why shall we not raise our voice against such type of culture? Those who have worked hard for the building of our nation, if their condition will be such, then who will get the respect? That type of culture has ruined our social structure.

Here I will give another type of instance where our culture is harmed seriously. In every year, our freedom fighters are honored in Independence Day, Republic Day, Kranti Divas etc. They narrate their past experiences, that how the country was made free etc. They complain that, their dream was something different. They dreamt 'Rama Rajya', but instead of that, they got 'Ravana Rajya'. Before independence, we were exploited by one, but now we are exploited by many, everywhere inside our country. The difference is that, we were exploited by foreign people, but at present, we are exploited by our people. The corruption is so deep rooted in the society that, we are feeling helpless. We are neglected in the society. Only on those days, we remember them. We are free without knowing the essence of freedom. So we are saying that our ancestors were tortured severely by the Britishers and tolerated so much of pain to make our country free. But in reality, we are not paying attention to take care of them, only saying so many things .This is only due to mechanical way of thinking.

Now, I want to cite another instance, where our previous system was better than present. We see corruption and pollution everywhere. Even we do not get pure food to eat. The food, that we eat are hybrid in nature. So, the original taste is gone. Our main food grain is paddy. Due to invention of various types of hybrid varieties, the native varieties are driven out which were very much tasty. Hybrid varieties produce more; farmers are interested to produce such type of paddy, in order to yield more. So also in case of vegetables, different kinds of hybrid varieties are invented which give more production. But these hybrid varieties require chemical fertilizers and pesticides in order to give more yields. These chemical particles are poisonous and remain in the grains and vegetables, which create different health hazards after entering into living bodies. Not only that, these particles also create problem for environment and ecology.

In ancient time, our fore fathers were farming with bio manures, composts; which were not bad for either human beings or domestic animals and products are tasteful also. They did not use poisonous particles in the field. So it had no bad effect on environment. But, now with the large scale use of inorganic manures and pesticides; the earthworms, birds and some helpful insects and bees which are helpful for pollination are extinct. On the other hand, the insects, though controlled to some extent, but gradually become accustomed with the poison and appear in a perverted form, which require hard poison to control them by bringing more bad effects to the beneficiary creatures. Our domestic animals, like; cows, buffaloes, goats etc eat the contaminated grasses, which are produced by the side of the agricultural field. So the poisonous particles mix with animal protein and enter into human body. Such type of situation arises due to farming of hybrid variety of food again.

Again, the milk that, God has given to the mother for the nourishment of her baby, is not free from pollution. Because we feed different types of chemicals in the feed to get more milk from cows and buffaloes. Sometimes, they are injected a type of medicine to produce more milk, which is very much unethical. Though we get more milk by adopting different types of chemicals or so, but it loses its original taste and creates different kinds of health hazards, which are not congenial to the environment. These cattle suffer from various types of diseases, whose treatment is costly also. But our native varieties of cows are very much befitting to our environment. Though they produce less amount of milk, these are very much tasteful and do not create health hazards.

If we go to our musical field, like song, dance, music etc, we can mark tremendous change in every field. Traditional song, music, dance, story are outdated after entrance of western culture. Once upon a time, the Ramayana was played in a stage. There was a scene like this, after marriage of Sakuntala, she was going to her father-in-law's house with her husband. But while she was leaving her father's house, she was weeping. There were so many spectators; both Indian and English. An English lady asked out of curiosity that, she is going to her husband's house, why is she weeping? An Indian, who was sitting by her side told that, this is our Indian family culture. When a girl goes to her father-in-law's house after marriage, she has to leave so many things like; parents, brothers and sisters, her native home, relatives, neighbors, village etc, which were familiar to her till date. How can she forget that attachment? Therefore, she is weeping. She is going to a totally new place, where everything is totally new to her. Then the English woman understood the Indian family system that how tightly it is knitted. But now such types of plays are rarely found, rather absurd and nude plays are more appreciated.

Likewise, our classical songs are rarely found now-a-days. Modern people have no interest for classical song, rather more interest for modern songs, which have no inherent meaning. On the other hand, they have dual meaning, i.e.; full with nudity, absurdity. We have forgotten our ancestors like Tansen, who was the legend in the song during the reign of Akbar. It is said that, he could bring rain by singing "Megha Mallahar". If our past culture of song was so much glorious, then why shall we forget it in the name of modernity? To-day songs are composed in such a perverted form that, conscious people, different organizations are complaining these as antagonistic to our culture. Rules should be framed to have control over these. The young people are blindly purchasing their cassettes without thinking the consequences upon the society. Sometimes, their songs, which have dual meaning are creating law and order problem in the society.

Likewise, in the field of dance, we see great change. There are different types of classical dances, performed in different states of India; such as, Odishi, Kathak, Kathakali, Kuchipuidi, Bhangda, Bihu etc. Apart from it, different tribes also have their own dance. These dances are performed with suitable songs, which carry certain meaning and impart some lesson to the society. These can be enjoyed with family, as there is no absurdity. So, in spite of our rich cultural heritage, we are following western dance, music and song etc, which pollute our environment and culture by bringing various types of criminal activities to our society.

So also our tales and stories, which are prevalent in the society, are gradually forgotten and confined in the books only. Our children are reading the stories of other society and culture, which are not easy to grasp on their part. Only they learn it mechanically. It has no impact on them. Here, we have to remember that, Shivajee became the great warriors by hearing the stories of great warrior from his mother, from the beginning of his childhood. There are so many stories in our culture which are filled with moral value. Again, as these were originated in our culture; the characters and incidents are easy to grasp on the part of our children. They hear these stories from nears and dears frequently, so it is become easy to remember on their part.

The children's mind is like clay, we can give any shape. So the stories were formed to impart morality through oral prescriptions among the future generations to create good citizen as well as good nation. Therefore, Indians are praised worldwide for their morality. Our two great epics The Ramayana and The Mahabharata carry so many valuable stories, we need not look anywhere. In ancient time, these two books were read in literate families in our villages at the evening and different people were hearing it sitting by the side. They remember by hearing and share their feelings among themselves as well as to their near and dear ones. As these were read in the families, the mothers were able to hear and tell their children while they go to sleep. As the mother is the first teacher of a child, these stories had great impact on the child and the future citizen is built accordingly.

But, if we look into the present society, such types of tradition are not found. The present generation is more literate, educated but morally deteriorated. In spite of all sorts of physical development, we are feeling suffocated. We fell as if we are lacking something. If it will continue, the society cannot be full-fledged. Our children cannot know our past culture. If our children cannot know about our past culture, then how can they respect our ancestors? Unless we know about different noble works done by Sri Ramachandra for his subjects, how can we respect him as ideal king? Therefore, the fault lies not with the present generation, but with the present system of education, which are framed by us for our children. Our children read the stories of foreign countries and work hard to get these by heart, because they are totally new to them. Afterwards, their behaviors are changed accordingly and commit different types of mischief, which are going in the foreign country.

In the name of modernity, our children follow the western culture from their school period. They follow the western culture in every respect; like, wearing of clothes, talking, behavior etc. and feel proud of it. They are also praised in their friend circle with the impact of foreign culture, the grown up children from different gender group create friendship among them and commit different types of nuisance activities, which are not expected from a student. Even a girl student becomes mother or commits abortion, which was not found in our society previously. This is a serious offence on the part of a student. When we open newspaper, such types of crimes are found here and there. Some wicked persons of the society take the advantage of it. So the western influence on our students has ruined the environment of our Indian student life, which was considered as the period of pure meditation. After looking the bad effects of western culture on our students, we are saying that, our educational system was good and congenial to the environment. The flux of the western culture has destroyed our young generation.

THE IMPACT OF MOBILE PHONE AND INTERNET ON STUDENTS:

The impact of mobile phone and use of internet have so many negative effects on student life. Instead of it's proper use, they use it improperly out of curiosity. Now-a-days, modern families are nuclear families, where both the spouses are office going and their children get sufficient time to use internet. It is also very much accessible. It is not easy to get sexual material from stores or to be engaged in sexual interactions under typical circumstances. Internet is not still photography; it can provide sound to moving pictures. Computers and internet provide us with virtual reality, which researchers have not yet examined closely. Internet is vast. The sexual materials on the internet are often free or inexpensive and are conveniently available twenty four hours a day. It is possible by the internet to disseminate enormous amount of materials on an international scale.

So people can get larger amount of pornography than ever before. The accessibility and vastness of internet paves the way for addiction to it. Some researchers suggest that, people can become more addicted to pornography, when there is vast use of internet. If more sexual and pornographic materials are available on the internet and more opportunities for interactions are possible, it will be reasonable to suggest that, some people will be spending more time engaging in compulsive sexual behavior. Again internet is interactive. Unlike other mass media, internet permits people to interact with many people under anonymous condition. Users can pay on-line services and that with sex workers or sex partners. Therefore, it is debatable that, whether internet is a blessing or curse? The internet can be seen as liberating for people with limited time. As with phone sex, it permits people to express themselves in ways that, they otherwise could not in a sexually safe environment. It is also evident that, shy people from intense and positive relationship on the internet and that some of these relationships become real life relationships. Some researchers say that, some combinations of anonymity and rewarding intimacy can lead to deviant sexual misconduct and disruption in material relation.

The internet is also difficult to regulate. In comparison with pornography purchased or rented from stores, it is difficult to prevent minors from accessing adult oriented material over the internet. Regulation becomes particularly a pertinent issue, when it comes to types of pornography that are less tolerated and illegal. It is mostly applied to child pornography, which is reported to be easier to obtain in the age of the internet. Any regulation is difficult in as much as the technology develops rapidly and the parties involved can become adept at circumventing regulation. Internet embodies many of today's latest development in media technology. Many of the issues, that crop up with any developing technology are surfacing with internet use, particularly sexuality. In as much as the Internet is in the evolutionary continuum of technological advancement, sexuality as with all human interactions on the Internet has evolved as well. As technology changes; social interactions, sexuality included will change.

Here, I am not debarring our children from the use of internet. There are vast opportunities for students to learn world class education. One cannot compete with others, unless he uses internet. So use of internet should be in the right way, i.e.; knowledge oriented, which will culture friendly.

Likewise, the use of mobile phone has both good and bad effects. One can contact his near and dear at the time of distress. We can take telemedicine and take the appointment of an important medical practitioner, who is staying at a remote place etc. On the other hand, criminals commit crimes and escape by contacting their partners. Boys and girls are committing different types of nuisance activities during their academic period. But the use of mobile phone has an important role in our daily life at present. We cannot deny the utility of it's service. Therefore, it should be used for the betterment of the society, for the benefit of the people. But in spite of all, the most important thing in this regard to be noted is that, the mobile phone, it's towers and internet nexus create health hazards and pollute environment which modern man has proved through research work.

ANIMAL FREEDOM AND MODERN CAPTIVE-CULTURE:

Another type of un ethical thinking growing among us, which is neither eco-friendly nor congenial to the environment, that, we are rearing hens, pigs and other animals in farms in unhealthy circumstances in large scale for making huge profit. We have to keep in mind that, we all are the creation of the same creator. Therefore, we all have the right to live freely and peacefully in the nature. No one should cross the border of itself. But, we the human beings are always crossing the border of every species for various human purposes and disturbing their tranquility. We are killing birds and animals for meat and other purposes and keep different types of herbivorous and carnivorous animals, like elephants, birds, deer, tigers, lions, bears, snakes, different types of birds and different types of aquatic animals etc. in zoo and aquarium under severe mental pressure for our amusement.

When a human being is remaining under mental torture; human right commission and other social organizations come to his rescue. In a bonded cage, whatever we feed to the animals, they cannot get pleasure. Their pleasure lies in freedom, in free movement in the nature. So the animals those who are wondering freely in the jungle are healthier than those who are in the zoo. Rather the wild animals, which are kept in the zoo in iron boundary, suffer from different types of diseases and ultimately die in spite of medical treatment. Previously, wild animals were used in circus, to show different types of plays to the visitors. But, now government has banned the use of wild animals in circus and also the domestication of wild animals.

On the other hand, when we are rearing animals in farms for meat, it seems very much unethical, because when we rear somebody, one type of affection come to our mind about that animal. Though a number of animals are reared in a farm, out of those, some animals those who are healthy and attractive may draw our love and affection. When those will be sold with other animals, definitely it would shock our mind. Again, when hens, pigs and cows are in farms, they remain in unhealthy circumstances. The animals, which are stronger and aggressive in nature, oppress the weaker, like take their feed, attack frequently, when they like etc. The freedoms of weaker animals are in danger and live always in fearful atmosphere.

So, due to different types of torture; they cannot flourish, rather suffer from various types of diseases, which contaminate other healthy animals and ultimately enter into human race. Bird flu, Swine flu is not human borne diseases. Bird flu is from hens and Swine flu is from pigs. As these are viral diseases, it could easily affect the other animals. I have seen that, when bird flu was severe among chickens in our state, the state government ordered to kill the chickens and warned the people not to take any more. So lakhs of chickens were killed and buried. But in spite of that, some might be thrown in the open field, which polluted the atmosphere and contaminated other birds which are flying in the nature, like crow, who might have taken their meat. As a result, thousand of crows were died, which keep the atmosphere clean by taking rotten meat and other rotten things in the nature and make the environment free from pollution. Likewise, swine flu created havoc all over the world. Many people died due to diagonal problem. Though it was originated in European countries but affected the whole world creating mega pollution. Our government took different preventive measures to control it, like medical checkup of passengers coming from affected countries through airplane, but could not control it. Many people of our country died out of it. So also AIDS is not a human borne disease. It was originated from monkeys of Africa and then entered into human race. Such type of diseases are pervading like epidemics. Though human beings have controlled the previous epidemics, like cholera, small pox etc. but new type of epidemics, such as AIDS, Bird flu, Swine flu are appearing and taking human lives in greater number. AIDS was originated from monkeys; but how did it enter into human society? The nature of disease is such that, it cannot be contaminated unless it comes in contact with blood or keep physical relation with the affected. Though there is no definite answer to it, but it is plausible that somebody might have close contact with the affected monkey.

On the other hand, Bird flu and Swine flu are air borne diseases. When human beings or animals become nearer to the affected animals, they are also affected. Hens and pigs, remaining in unhealthy circumstances breed such type of diseases and contaminate others. These diseases were not found before. The animals were living and moving in nature freely. When human beings kept them in a confined area, pollution started. Such type of culture was generated in the developed countries. As they are technically developed, they could control the pollution created out of it. But, we the developing countries, though follow their path, cannot control the pollution created out of captive farming. Once pollution started means, it will spread to different parts and the environment of a larger area will be polluted. Healthy animals and human beings will die in large number.

MIS-INTERPRETATION OF RELIGION IS A THREAT TO THE HUMAN RACE:

Rich and developed countries sell arms and ammunitions to the developing countries and get large amount of money. The third world countries are not ready to understand their own problems like health, education, food etc. rather involved in quarreling with neighboring countries. Pakistan is the best example of it, who is always trying to create disturbances in our country by training the terrorists. Kasab, the Pakistani terrorist of 26.11.2008 Mumbai attack is the burning example of it. He was educated up to class-IV and left education due to acute poverty according to the confession made by him. After that, he was involved in burglary and finally joined in a terrorist group.

Here, a question arises, if the national leaders of that country would think for the development of their people, such type of problems would not arise. Rather, they use the foreign aids for purchase of arms and ammunitions, to use against our country. If they would use the foreign aids for the purpose it is given to them, then their human resource would be developed and ultimately it would be a benefit to their country. They are creating bloodshed, destruction of private and public properties in the name of Jehad. But no religion says so. The essence of every religion is for the construction of mankind, not for the destruction of mankind. They are treating terrorist attacks as Jehad, which is great threat to the mankind as well as to the environment and ecosystem. Here the word Jehad is misinterpreted. Where human resource is developed, no such idea will come into their mind. Religion is framed for keeping up of people, not for killing up of people. Only some fundamentalists are motivating people for committing such type of nuisance activities. They are sending terrorists into our country in disguise, to crate disturbances in our country, like attack on different shrines, Parliament and different important places, where great loss can be incurred to our country.

So, in the name of religion, some people are inculcating poisonous seeds among the innocent people, exploit their emotion, cause loot, murder, massacre and social setup is imbalanced by destroying the environment and eco-system which are created from centuries long.

PECULIAR PROBLEMS ARISING OUT OF INDUSTRIALISATION:

When a mega project is undertaken by government or private companies, they face so many obstacles from different sides, such as:

Political, Regional, Religious, Environmental, Economical and Cultural.

When government tries to set up a project, the opposition parties showing different pros and cons motivate people to protest it. The chief intention is like this; if the government will establish the industry, the credit will go to the ruling party, which will be helpful to win in the next election. They project the minor causes as major and bias the people.

Regional cause is another cause, which obstruct the establishment of industry. The regional and local people of our countryside are so orthodox that, they cannot accept any change. Though they go on starvation, but reluctant to accept any kind of change. If industry will be set up, different types of people will come to our locality, they will share our resources. The peace and harmony of our locality will hamper etc. When industry will be set up, different types of anti social activities, like *Gehero, Bandh*, and Labor problem will arise side by side. When such type of activities will go in the locality; the local politics, local people will be involved, which will affect the peace loving and innocent people of the locality. Therefore, the local or regional people oppose to set up industry in their locality.

Religious cause is another cause, which obstruct the establishment of industry. If industry is going to be set up near the shrine or by evicting the shrine, it will face serious challenges from the side of the particular community. During the period of BALCO movement, a temple of lord Shiva was a cause of movement, which was said to be slightly affected or so. People joined other causes with it and tried to catch the sentiment of others in order to strengthen their movement.

Environmental cause is the most important cause, which is created out of set up of industry. People are very much aware about the environmental problems. Though government has created various departments for check and control of environmental pollution, caused due to set up of industries, the corrupt and dishonest officials give permission for set up of industry by taking bribe. But the local people revolt against it and industries cannot be set up.

Here a very simple thing, which I have marked in jungle areas; where forest guards are watching for the safe guard of the jungle, jungle is depleting gradually. But when the local people get involved for the safe guard of the jungle, the density of the jungle has increased gradually. So, environmental factor is very much crucial for the establishment of industry. On the other hand, the industrialists are rich persons; they live in metropolitan cities and maintain very much sophisticated lives. They are not affected due to establishment of industry; even they do not know what problem the local people face due to set up of industry. In Bhopal gas tragedy, thousands of human lives were destroyed before the sun rise, but the chairman of Union Carbide Company, Mr. Anderson was snoring in deep sleep in remote place. So, experiencing different huddles, created by industries, people are opposing for set up of new industry.

On the other hand, those industries which use non-renewable resources, as their raw materials will be complete after few years, they face various challenges from the side of environmentalists, public etc. the technology has so developed that, if there will be no checks and balances for the use of resources, then those will be utilized within few years only. Our future generation will not get to use. So the theory of sustainable use is adopted for natural resources, keeping in view of future use. There is a saying that:

"We conserve only, what we love.

We love only, what we appreciate.

We appreciate only, what we know.

We know only, what we are educated."

Industries face so many problems from economic point of view of the people. Indian economy is chiefly based on agriculture. People, who live on agriculture, also depend on forest, i.e.; for making ploughs, fuel woods and other house hold purposes etc. The tribal people live, cultivate and collect forest products for their maintenance etc. When a project is undertaken to be set up, mostly it requires such type of land. The people, those who are benefited out of it, protest against the project. It is said that, life and livelihood are on the same line. So people become united for the safety of their native land. If the cultivable land is fertile and multi-cropped, then the protest is severe, which is going in proposed POSCO site of Odisha . As that is highly fertile land, the local people of that area are opposing vehemently. Even if they are assured to get higher amount of compensation, still they are not ready to leave it, because they are acquainted with such type of economy on ancestral basis.

Culture is another factor, for which the local people do not agree to leave their native places to set up industry. Mother land has a special kind of attraction for every human being. Because, he is born here and will be cremated in the same ground with his predecessors. The people, those who live in a place, create a certain type of culture, which binds them together in spite of several differences. They observe the national, social and communal festivals united. During the time of any kind of disaster, they join their hands with each other and mitigate their problems. Temples, Churches, Mosques; which are situated there will be ruined, if project will be undertaken in that locality. Even though they are assured for rehabilitation and resettlement, still they are uncertain that, they will get back their harmony. Because there is no guarantee that they will get same type of neighbor. Their nativity will be disrupted. All such things came to the mind of the people and they do not agree to spare their native land for project work.

Therefore, government should frame policies for rehabilitation and resettlement of the displaced persons, who are ousted due to project work. The displaced persons of Hirakud Dam of Odisha have not got their compensation till now, neither in kind nor in cash, so also in Paradeep, NALCO site. So people are opposing for new projects. If the local people, NGOs, local leaders (both political and non political), educationists would be involved before setting up of the project, then the project work can be done. We know that, economic development cannot be possible without setting up of new industries. But it will be more profitable, if these industries would be based on renewable resources like agricultural products or so, where agriculture and industry can go hand in hand. Those industries, which are using traditional energy and producing more heat and emitting heavy chimney gases, should be controlled, because they are polluting the environment, polluting the earth by making hotter and hotter.

VARIOUS REASONS FOR INCREASE OF TEMPERATURE AND IT'S EFFECTS:

When we say about the environment; the increase of temperature and the average temperature of last ten years are to be taken into account. According to the historical evidence, the temperature of the world varies from time to time. The world temperature increased from 1915 to 1940 and then decreased up to 1960. After that, temperature is increasing year by year. An advanced scientific organization of America 'National Academy of Science' has shown in it's report that, the temperature of world has increased by 0.6 degree Celsius within last century. Though this is not so much, but in the last twenty years, the rate of increasing is alarming.

If the increasing tendency of heat will continue, then it will be 1.1 by 2005, 3.0 degree up to the middle of this century, i.e.; 2050 and at the end of this century, it would increase by 4.5 degree Celsius. It is calculated that, within 125 years, eleven hottest years occurred in last twelve years. There are two causes for change of temperature. One is natural and another is artificial or man-made. Natural causes are continental movement, volcanic eruption, oceanic current, comet and other heavenly bodies. Man made causes are excessive growth of industries after industrial revolution and luxurious life style etc. In other words, the heavy growth of carbon gases in the atmosphere and depletion of forests are the main causes. The presence of gases in atmosphere is expressed in PPM. Before industrial revolution, the presence of carbon dioxide gas in the atmosphere was 280 PPM. In 1986, it reached to 340 PPM, in 2005 it reached to 383 PPM and now it is going to be 400 PPM. Every year it is increasing by 2 PPM. But the safety limit is 350 PPM, which we have already crossed. The felling down of forest in the name of deforestation is another cause. According to the data supplied by Food and Agriculture organization of UNO, the total area of forest is 3.4 billion hectors, out of which 52% are found in tropical zone. The destruction of forest in that area is the main cause of global warming.

The forest acts as natural sink. Every year, we are destroying 120 lakhs hectors of forests. Therefore 100 to 120 crores tons of carbon are entering in to atmosphere without being absorbed by the natural process. It is more in the developed countries than in the developing countries. An Indian exits 1.2 tons of carbon every year, where an American exits 21 tons of carbon every year. Likewise, if we take the use of energy as measure of development and if the use of energy of an Indian is the base, then a Chinese use 2.5 times more, Japanese 13 times more and an American 26 times more.

Now let us see, if the temperature will rise, what will be the results? If temperature will increase, glaciers will be melted. Though scientists say that, there is a natural cycle, which acts behind the melting of glacier. But no one can deny that, the frozen ice of Antarctica has started melting, which were heaped hundred and hundred years before. The scientists of NASA have stated that, the ice of Green Land started melting. So the water level of ocean will increase. There are 15,000 frozen ice pieces in the world, out of which 9,575 pieces in our country.

When glaciers will be melted, rivers like Ganga, Jamuna, Bramhaputra and Indus will be affected. Those navigable rivers will dry in summer. The cultivation work in Ganges plane, which is the food basket of our country, will be affected severely. The water got from dews will be reduced, due to increase of heat. The humidity of our country will be reduced up to 5% for excessive evaporation of water. When this will happen, the area where rain fall is low will be tuned into barren land. Worldwide life organization has shown in his report that 3 million hectors of land are covered with glaciers, where 12,000 cubic kilometer drinking water is stored. If these will be melted, people will face the problem of drinking water. According to the view of Natural Water Commission, 1820 cubic meter water was available for individual in 2001 and it will be reduced to 1000 cubic meter in 2015. Due to scarcity of water, people will face problem in the world.

It is estimated by the scientists that, the average temperature of earth will increase 1.8 degree to 4.8 degree Celsius and the sea level will rise from 0.6 meter to 2.9 meter. Two causes are responsible for it. One is the melting of glaciers and another is expansion of oceanic water for increase of temperature. Therefore, the sea level will definitely increase; as a result, different populated countries like Japan, Sri Lanka, Thailand, Indonesia, Myanmar, Singapore, West Indies, England, Holland, France, and Portugal will be affected. America cannot be left out of it. Our Mumbai, Goa, Andaman, Chennai, Kolkata will be drowned.

The increase of temperature has great impact on weather. Hot climate affects the natural rainfall as well as snow fall. Devastating rain, drought, small rain fall are the causes of famine. Storm and cyclone are the result of the global warming. We all have experienced it more or less during previous years. Now-a-days more rain fall is occurring due to rain depression, rather than the impact of monsoon and the rate is increasing in Bangladesh and India.

The flora and fauna which are grown in the natural habitat will be affected due to change of climate. When sea level will increase, it will drown the low lands of coastal belt and destroy different species of trees. The sphere of biodiversity will be reduced. The productive capacity of trees will be reduced, due to excessive heat. Especially, the paddy and wheat of the tropical zone will be affected. Though it is right that, the crops grow quickly due to more heat, but if it will be more, the process will be expedited. It will get less time for storage of carbohydrate, for which production will be less and the amount of protein will be less also. The great Pandas of China, a rare class bear of Yellow Stone Park in America, the tigers of different reserve forests in India will be badly affected. Due to climatic influences on seasons, the annual migration of birds will be affected.

Again, if one degree Celsius will increase in temperature of the world, then ten crores of people in our country will suffer from various diseases. Mosquitoes and various germs of diseases will multiply their families and Malaria, Dengu, Chickengunia will attack people. The health of human being will decrease due to epidemic, malnutrition and dehydration etc.

Though, many scientists do not hold the same opinion, regarding the bad effects of global warming, but it is sure that the temperature of the world is increasing gradually. Suppose a frog is in the water tob and the water is gradually heating, but the frog cannot know and he is becoming adjusted with the temperature. But when the water will start to boil, what will be the fate of the frog is known to everybody. Likewise, the sea level is gradually rising. Who knows that, it's reflection will be not in the form of Tsunami or Tornado. At that time, sea shore villages would be drowned. A new type of rehabilitation problem will be created, due to natural calamity. Shortly, it can be stated that, the increase of temperature will ruin our fertile land, drinking water, biodiversity etc.[12]

As explained above, we find different reasons and results of global warming, which is a great challenge before the humanity. In order to face the intricate problem, man should be well prepared, especially; he should inculcate a social culture in which he can avoid global warming, or else, there is no rescue out of it.

---

12. Odia Daily News Paper 'The Samaj' dt. 14.10.2010

IRRATIONAL USE OF WATER CALLS FOR VARIOUS PROBLEMS:

All the living beings depend on water and environment for their very existence. But now, there is crisis in these two fields. We are misusing water for our own interest, without thinking about the consequence. It is said; if there will be third world war that will be for water. Economics has divided goods as free and economic. The water in the sea, river, stream, and pond is free goods. But when we purify it or extract underground water through pumps, it becomes economic goods.

Due to excessive human use, the water is becoming scarce and time will come, when we cannot get water by spending money also. As we are not careful in using water, there will be crisis in future. An international seminar was convened at Chandigarh in 2009 regarding scarcity of water. The report was published like this; the patients, those who are under treatment in different hospitals worldwide, 50% of them are suffering from water borne diseases. The use of contaminated water is the chief cause of it.

Again it was stated that, India would face acute scarcity of water at 2025. In our country 14% pipe water are wasted in municipal areas. If we will take into account our state, the water of some rivers like Bramhani, Mahanadi, Baitarani, Subarnarekha, Rusikulya are not fit for use. The effluents of industries have polluted the rivers. Not only men, aquatic animals, especially fishes are in danger. Media is publishing regarding the unnatural death of the fishes in different rivers in several times, due to contaminated water. Though there are rules and regulations, but companies are violating frequently. As a result, problems are arising in drinking water, agriculture, aquaculture and use of water in other domestic fields. It is known that, crores of rupees are pending with companies as due, those companies who are taking water from Hirakud Dam. Government is not emphasizing for collection of those dues from them, rather pressing the common people and farmers to deposit water tax.

Our state is facing water crisis since two decades. It has become acute not only in hilly areas, but also in coastal regions. It is granted that, if one liter of water contains 1.5 mg fluoride, then it is fit for use, but more than that is dangerous to health. But it is examined that, water in about 15 districts contain more than the recommended amount of fluoride, which is not fit for use. People are complaining the government agencies, arranging rallies, picketing, striking etc., but in vain. Here, one thing we have to mark that, not only government is showing favour for establishment of industries, but people are adding manures in ponds for growth of fishes, building houses by filling sweet water areas etc., which go against the safety of the sweet water. I have read from a news paper that, people died in diarrhea in the district of Rayagada, by using contaminated water. On the other hand, companies are violating the terms and conditions, made by the government for use of water. Even if, they are dragging more water from river and underground according to their need without caring for government contract. Tanks and wells are becoming dry due to such type of activities. They have become accustomed with such type of activities and not responding the warnings given by the government.

So problems cannot be solved by blaming the government only. We have to become conscious and lack of consciousness is the cause of water crisis in that locality. Water should be used sustainably, keeping in view of other's necessities. But the thing is going in opposite way. We have to keep in mind that, the usable water is not unlimited. If we will not be conscious, in future, there will be struggle among the people. Everybody should be careful about it, both at government and pubic level. People should be made aware through different slogans, meetings; banners etc. or else our future life will be full with sorrows.[13]

Here, the thing to be noted is that, the water crisis which creates environmental problems is due to cross cultural thinking. We are cultivating such a culture, in which there will be water-crisis. Boring system of water collection is such a culture in which we drag huge quantity of water and use for agriculture and industry and the water level is going deeper, for which some state governments have banned this boring system. In the primitive age, main water collection was done by ponds and tanks, which could protect the water level. Our forefathers worshipped the water as Baruna, realizing the

---

13. The daily news Paper' The Dharitri' dt. 01.12.2010

importance of water. However, we may promote such a culture, in which, we can avoid water crisis or else, it will have wider impact on environment.

ABNORMAL SYSTEM OF BIRTH GIVES RISE TO DESTRUCTION OF ECOLOGY:

Development is a continual process, no one can check it. But the intellect persons always pioneer in the progress of the world. Different types of scientific inventions take place for the betterment of the society. New knowledge is obtained in various fields like medicine, surgery, communication, space research etc. Medical science has invented various tools and technologies for support of lives. In Toto, death rate is decreased, birth rate is controlled by applying various devices and means. Now human beings are thinking to be immortal and searching the cause and have found the DNA, which is responsible for becoming old. It has been invented that, the lady who is issueless, incapable of producing child and has great anxiety to become mother will be successful in her mission, if the couple will accept the advice of the expert doctor.

The doctor will collect the sperm and egg from the couple and fertilize in the laboratory and put it in the womb of another lady, who is willing for the process. She will bear the baby till the birth. The couple will take care of that lady and pay the amount according to her demand. She is called surrogate mother of the baby.

Here the work is noble. Because childless couple can be blessed with a child, so also love, affection and the family will be overwhelmed with joy. But another thing is growing side by side, that is in rich and ultramodern families, the females are not interested to be mother. They arrange a lady, who can carry their baby in exchange of heavy amount of money. The psychology that works behind it is that, if she will be mother, her beauty will be lost, she will take so much pain etc. Here, so many questions come to our mind. When a child will be born with the flesh and blood, of another woman, definitely the child will have more affection towards her. Though the child is handed over to the original parent, just after the birth, still in course of time when the child will be grown up, will know everything from family, friends, neighbors etc. At that time, the love, affection what he has in his mind to his parents will be lessened and love that surrogate mother who has given birth. Here the value of the word 'mother' is lessened due to such type of activity. We have made human beings child producing machine. This is due to our mechanical thinking and we have made lives machines.

Not only surrogate mother ship, but also in the field of cloning, we find out a great devastation to the future generation. Imagine, if the cloning system will be rampant, what will be the condition of environment? The total system will be collapsed due to introduction of cloning culture. It is for this reason, that the total system has been nipped in the bud all over the world.

REVIEW:

As is evident from the above discussions, we find that we think in a different line that is vitalistic or value-oriented, while the question of environment and ecology comes. Again, we act in a totally opposite line, while the question of environment and ecology comes. We always act on mechanistic aptitude in this regard. We are in need of a culture, which will protect the environment and ecology. But we are adopting such a culture, which will destroy the environment and ecology.

The whole thing of environmental and ecological hazard is mainly due to our adoption of mechanical culture, in which there is little consideration of value-system and maximum consideration of monetary gain, where production, distribution and monetary profits play the major role. If this will be our tendency, then we cannot safeguard the interest of environment and ecology. The world will see great destruction and entire civilization will collapse like a house of cards.

****

CHAPTER-IV

PHILOSOPHICAL THINKING OF MAN IN TERMS OF ENVIRONMENT AND ECO-SYSTEM

PHILOSOPHICAL THINKING OF MAN IN TERMS OF ENVIRONMENT AND ECO-SYSTEM

Innumerable species are seen on this earth, which have a good number of diversities and peculiarities. The earth is so beautiful and the varieties are so large that, when a man sees it and thinks over it, he becomes a philosopher. Almost every man has that insight to think philosophically on the diverse phenomena of this planet. That is why, there is a common saying:- "Every man is a Philosopher". Why every man is a philosopher? It is because of the fact that, man has that propensity to think rationally and philosophically, to think in an innovative way, to proceed in a rational manner and to rectify it's rational thinking according to the needs of time and circumstances, to innovate some ethical standards and value system, to protect all the species of our beautiful earth and to protect the entire cosmos and to do something beautiful in this regard.

Other species, although they are conscious, yet, their consciousness is very low where there is no scope to think in a better manner and to take the leadership, to do something constructive for the promotion of environment and ecology. It is only man, who can take the leadership and present a philosophical thinking, to protect environment and ecology. Nature has placed man in the highest position for the promotion of ecology.

But it is not a fact that, other species are violating and destroying the environment and ecology on this earth. Rather if we think deeply, we see that, other species follow the law of nature, which is most vital for the promotion of environment and ecology. For example, a tiger resorts to sex once in a year. We do not find any violation of the nature in this regard. Just like that, every other species, besides man follow the law of nature in a perfect tune, which is essential for the protection of environment and ecology.

As we said before, although man is the fittest species to think philosophically and to protect the environment and ecology, yet sometimes, it's activity and thinking goes directly against the aforesaid constructive and value oriented notion. Sometimes man becomes so dangerous that, it brings great loss to the creation, which the other species cannot do. It is because of the fact that, God has given enough freedom to do the man, who is to utilize this freedom in a constructive way. If man uses it's freedom in a constructive way, then it becomes conducive for environment and ecology. If man utilizes it's freedom in a destructive way, then it becomes most dangerous for the promotion of ecology. Man has the freedom to obey and violate the law of nature. It is for the man to take up the first alternative and to proceed ahead, taking the leadership to protect the environment and ecology.

THE IMPORTANCE OF FOREST:

When we say about environment and ecology, first of all, the idea of forest comes to our mind.

Just after independence of India, there was heavy demand on forest land for development. Rightly then the programme of "Vanamahotsav" was lunched with all sincerity of purpose. The celebration of the Vanamahotsav week has become more ritualistic. Since 1872-Stockholm-Conference, the importance of forest has been felt by all concerned. But the pious intentions have not been able to retard the tide of growing enormous pressure on forest from human and cattle growth added to rural poverty, agriculture, multipurpose dam project, industrialization and consequent pollution, rehabilitation of displaced villagers etc. Forests devoured with demonic menace. Due to these demands considerable forest areas are lost. It is reduced from more than one third to less than one fourth geographic area of the country's land mass. There has been eternal clash between supply and demand and protection and plunder. The greed outgrows need when it comes to encasing from open treasury, a natural renewable resource. The forest areas are reduced and fragmented causing serious threat to the environment and wild life habitats, for which they are gravely imperiled. Man becomes the predator in this uneven and continuous fight for survival of forest. The forests have more sublime and spiritual dimension along with social and economic functions.

Forest is a national asset and has to be managed sustainably for a fair dividend harvest. Regeneration and plantation are essential for management and development of forest. The livelihood issues of tribals and people living in and around the forest have to be addressed as an important threat of forest management. The forest areas are reduced with biotic pressure along with other demands. Govt. launched social forestry, joint forest management to stop the depletion and develop forest for people. The wisdom for this effort came after the loss like: - *"Teri yaad aye, tere jane ke baad"*. Whatever the case may be, it is better late than never.

PROTECTION OF FORESTS BY THE GOVT AND NON-GOVT AGENCIES:

All heterogenetic and complex issues of forest have to be managed in a homogeneous and holistic manner. Priorities have to be set and work is to be done with a vision. *"Loss of forest is loss of civilization"*. In 2004, Ruskin Bond said the role of foresters "as they have given green growing land instead of desert land".[1] There are lots of misuses of forest land for agriculture, irrigation project and forests are the last item in the govt. agenda. As they have no voting power they are in the last protocol. But now the scene is changed. Govt. is deputing rank officers of forest department to foreign countries, rich with forestry to pursue important current issues like global warming, climate change, desertification, maintenance of biological diversity, carbon trading, clean development technology, sustainable development, wild life and zoo management etc. If we consider the facts of later period, I have heard that forest was protected and nurtured by innumerable Munis-Rishies in India and Ashram civilization was developed in and through the forest.

This was the Puranic period where protection of forest and nature was the inherent slogan of people. I have ever heard from my grandparents that when India was under the British rule, the local kings were ruling in hilly areas. They were taking the proper care of the forests. They were giving permission to the public during a certain period of the year to collect the fire wood without destroying the valuable trees like Sal,

---

1. Souvenir-2009, The Society of Retired Forest Officers, Odisha, Page-11.

Piasal, Sisam, etc. But maximum damage started to the forests after the independence. Even, corrupt forest officials joined their hands with Mafias. Forests form an important component of environment and it's impact is quite considerable. It is stated in state of India's environment- The second citizen's report 1985 "The environment is not just pretty trees and tigers …………….. threatened plants and eco-systems. It is literally the entity on which we all subsist and on which entire agriculture and industrial development depends."[2]

The union ministers of Environment and Forests call for a meeting of Forest Ministers from all states annually, which becomes a formal affair without any concrete solutions of problems. Govt. of India has several regional offices where state specific issues can be sorted out in quarterly and half yearly meeting. The suggestions from those meetings can be placed in annual Ministerial meetings to get a direction from the Govt. of India according to the state needs. The forests should be managed from beginning mostly for revenue earning by harvesting the surplus, recommended in working plan or technical document and the preparation of the document need time and detailed field inspection. The basic issues of maintaining health and hygienic of the forest, adequate forest inspection and account keeping should be adhered to. Now there are so many antisocial elements active in the forest areas. Field inspection has become risky in vulnerable forest areas. The stress is to maintain the territory of forest to get benefit of clean development. In jungle areas the Mafias are looting the timbers equipped with modern weapons. Staffs of the forest department have no such weapons, their number is very less. So they cannot protect the forest endangering their lives. Sometimes forest staffs are also preyed by them and loss their lives. Very often we find such type of news from different news papers. Again Maoists are remaining in the forest and conducting different types of activities inside the deep forest. These are some of the incidents which break the mentality of forest department officials to enter into the forest fearlessly.

---

2. Ibid-Page-11.

But, now-a-days the outlook has been changed. People living adjacent to the forests are getting involved for protection of the forest. Even if trained elephants are brought and engaged for watching of the forest. In Similipal, some trained elephants are engaged to save the trees from felling down by Mafias and thieves.

These elephants enter into deep forest where there is no road for vehicle. We have not shown any sincerity before cutting of forests. When we feel it's bad impacts, then our consciousness arose. We should look any issue with positive bent of mind and proactive manner. The present day trend and demand is radically different from the past. Ecology and Economy have to go hand in hand with adequate preservation of forest resource. Forests belong to whole country and no one has the monopoly to usurp it to himself. We should strive to conserve mange and sustainably develop forests for prosperity, for environmental and ecological security, the very survival of humanity.

FORESTS BRING HAPPINESS:

Once Mr. J.W. Nicholson, the Provincial Research Officer of Forest Department in 1925 had said, there is an old Hindu saying that "where there is no jungle there is no happiness.[3] The truth of this statement is exemplified in Bihar, Odisha, where the happiest and most contended races are invariably to be found in the forest tracts." At present only remnants of forests are quite insufficient to meet the demands of the population of one district. Steps were taken for conservation of forest during the period 1870 to 1880 in Bihar and Odisha. The people in general, were against the conservation of forests. The forest officials were victims of suspicion. Lack of knowledge about forest, apathy of people towards forest protection, indiscriminate cutting and clearing the forests were the problems in those days. In spite of these problems there was no environmental degradation which we feel more and more today. In Odisha, 37.34% of land area is forests area.[4] The total forest area is 58136.23 square kms. The record shows that an area of 287.68 sq. kms has since been diverted for non forestry use by different developmental project, mines and industries.

---

3. Ibid-Page-13
4. Ibid-Page-15

In spite of reorganization of forest division, the scene is not improved in maintenance of the forest boundaries, repair to the forest roads, post planting care to the plantations and adequate protection and illicit connection of forest officials in timber smuggling, poaching of wild animals and irregular mining in the forest areas. The forest fire, uncontrolled grazing in the forest area is common in almost all government forests of the state. The departmental statistics indicate the list of forest offences are in the increasing trend and invariably the forest officials shift their responsibilities when it comes to their efficiency or lack of interest in proper protection of forests. It is also fact that the political will is very much lacking and avoidable interference in the forest administration with threatening of transfer or vigilance raid prompt the sincere and honest officers to change their strategy of administration to suit the local need. Recently, the forest department has supplied mobile phones, VHP sets in the remote areas and organized gang patrolling. But the system needs to have more improvement while the public should be taken into confidence to combat the interference by the unscrupulous forest offenders. The participatory forest management miserably failed and need review before it is too late.

Forests are the promoters of environment. We should think globally and create consciousness of the public as well as officials for the security of forests. Meetings should be held frequently in order to increase co-ordination between forest officials and local public. We know that, the economic conditions of forest dwellers are very poor. So they should be given some facilities for use of forest without doing harm to the forest. If this will be done, forest areas will be increased and the loss, which have incurred to the environment will be compensated on gradual process.

DIFFERENT PLANTATION SCHEMES FOR CREATION OF FOREST:

The plantation of forest in the state has gone through many ups and downs also. Initially it was taken up by Forest Department and then by the plantation corporation which ultimately merged with the forest corporation till the plantation activities of the corporation was transferred to the department and finally now being taken up through the Forest Development Agency (FDA). Now Govt. of India has taken up two major schemes namely Revised Long Term Action Plan **(RLTAP)** and **NAP** under which plantation programme is going in the state apart from the plantations under the state plan schemes like Economic plantations, Jagannath Ban Prakalp and medicinal plantations.[5]

RLTAP (*Revised Long Term Action Plan*) it is a central scheme of revised long term action plan for KBK districts of our state. The main activities under this scheme are to reforest the degraded forests to support the rural poor, ensure the availability of the forest products to the local inhabitants to support the forest based livelihood and improve the forest cover, soil and moisture contents while generating opportunity for employment.

NAP (*National action plan*) It is also a central assistance implemented in association with Forest development agency (FDAs). The scheme started in 2001-02 and it will continue during the 10th plan period. Government of India will provide fund to the

Forest development agency (FDAs) directly for plantation in each forest division and all programmes will be implemented by VSS (Vana Surakshya Samiti).

Jagannath Bana Prakalp: It is an ongoing scheme of the state, which has started during 1998. It is said that 2095 hectors have been covered during these years with public participation of 618 divisions of the coastal region with an objective to meet the requirement of the car festival of Lord Jagannath at Puri in coming years. Funding for the scheme is met by the state budget. [6]

Economic Plantation: - This is a state plan programme and annually 3000 hectors are being taken up for plantation of economically valuable forest species, mostly in the reserve forests.

---

5. Ibid-Page-15
6. Ibid-Page-16

Medicinal Plantation: - This is a scheme being operated by Forest Department in collaboration with the State Medicinal Plant Board. In-situ and ex-situ conservation of the existing medicinal plant in the forest is taken care of while creating awareness among the public for conservation and propagation of the important medicinal plants in their locality. This scheme has been implemented in seven forest divisions of the state.

We should sponsor our co-operation for successful implementation of various plantation oriented government projects in order to increase the forest areas in our state, for which our future generation will be benefitted, our necessity will be fulfilled in the long run and at the same time environment and ecology will be balanced. Our motto should be substitution, not depletion.

FORESTS PROVIDE LIVELIHOOD TO FOREST DWELLERS:

As forest plays a key role in maintaining ecological stability, reducing impacts of natural calamities such as drought, flood and cyclones, recharging ground water levels, checking soil erosion and providing livelihood support to a large segment of population, government is taking different measures for growth, development and protection of forest areas. In Odisha 22.1% is Scheduled Tribes and 16.5% is scheduled castes of its population. Tribal communities have close cultural and economic ties to forests. About 67% of the population of sixty two Scheduled Tribes stays in constitutionally mandated Scheduled areas that cover more than 44% of the state land area (seven districts full and six districts in part)[7]

The tribal people worship forest as God and take care of forest. Even if they have been using forest on ancestral basis from generation to generation, no harm is done to it, in spite of sufficient use from different point of view, like food, fodder, housing materials etc. But when the civilized men interfered on the forest habitat, destruction was started. The problem of environmental degradation and ecological imbalance came to a remarkable point, where each and every conscious person was expressing his deep concern regarding how to save the environment.

---

7. Ibid-Page-17

Last but not least, the question now arises when tribal people use forest, there is no harm; and when the modern man uses forest, there is enormous harm. As to this question, the answer is that the tribals in their simplest form were using for their basic needs, but the modern man is now using for its greed. The world is enough for the need of man, how large its population may be; but the world is not enough for the greed of a single man only. Here lies the secret of the pollution of environment & ecology. These tribal districts mostly have higher forest cover than the state average. Both within and outside scheduled areas tribals live on farming, forest foraging and wage work. It is estimated that in 8.8% of forest area, shifting cultivation is conducted. Thirteen communities identified as primitive tribal group. They largely practise a pre-agricultural foraging way of life. The Odisha State Development Report characterizes the tribal economy as primary subsistence oriented and based upon a combination of agriculture, forestry and wage laborer.[8] Though tribals are mainly dependent on agriculture, collection from forests continues to play major role in house hold consumption and income generation. As forests are primarily the source of house hold consumption and income generation, mostly they do not prefer to accept agrarian livelihood system in those areas.

The literacy levels of SC/ST are very poor in Odisha. In 1991 it was assessed that the literacy level of ST men were 27.93%, SC men 43.03% as against 63.50% for other castes.Likewise, only 8.29% ST women and 17.03% SC women were literate against a rate of 39.54% for other castes.[9] This shows several alternative livelihood options for SC/STs, who would remain dependent on land/ forest/wage based substances in the foreseeable future. On the other hand, almost, half of the Odisha's population is poor. The percentage varies from 45 to 50% of the total population. This widespread poverty Is linked to low literacy, poor health status including higher infant mortality, limited irrigation development, low agricultural development and limited secondary sector

---

8. Ibid-Page-17
9. Ibid-Page-17

employment opportunity. In 2001, eighty five percent of Odisha's people live in villages. Out of them 26.11% are tribal. Therefore, an impoverished rural population is much more susceptible to income shocks and has greater dependence on forest for survival, food, raw materials etc.

ODISHA FORESTRY SECTOR; VISION 2020:

The Odisha forest Department in collaboration with DFID (Department for International Development) began the exercise of developing vision 2020 during April 2004 and has finalized it after consultations with all level of forest personnel, other departments and field visits.

Odisha's forests are well stocked, diverse, multistoried and dense. The forests are managed for sustained use providing a range of goods and service to a variety of stake holders at local and broader levels. Forests are home to flora and fauna, build soil, regulate water flows, provide quality water as well as small wood, timber etc. They provide inputs to the agriculture economy, income to the impoverished and subsistence livelihood to all. An enabled and responsive Forest Department and empowered local community, institutions collaborate to protect forest areas from encroachment, poaching, illicit felling and fire.Flexible forest tenure allows a diversity of institutions and forest categories at different spatial and temporal scale. Categories reflect function. Protected areas protect representative habitats and species. Reserve forests are for environmental protection, subsistence use and commercial production. Forests are managed under a range of participatory options, for local use, sale of surplus and commercial use. All categories however respect local rights and provide strong incentives for local participation in protection, management and use. Diversions of forest for development projects invite disincentives and increased cost and local compensation after due process. Transparent operations and procedures reduce transaction cost for harvesting timber and starting and operating enterprises, reduce externalities in terms of social and ecological impacts and maintain confidence in the integrity of forest management and operation.

The eleventh plan of our country emphasizes on reduction of accelerated poverty and higher economic growth with social justice. Therefore, the intention of the forestry sector is to promote sustainable forest management in the state with a larger goal of supporting rural livelihoods by recognizing the relationship between rural livelihoods and conservation of forest resources in a sustainable manner. Investment in forestry in the state will directly benefit the most vulnerable groups of the people who are the poorest among the poor. Therefore the first and foremost duty of the forest department is to protect the dense forests in the state with their biodiversity and wild life from degradation and to restore their natural regeneration through appropriate treatment. The degraded open forests must be arrested and these forests which are open must be regenerated and reforested to improve their crown density. The forests which have lost the indigenous rootstock must be regenerated through artificial and natural means. Concrete efforts are taken during the XI plan period to improve the economic value of the growing stock of these forests by taking up programme of reforestation with economic species. However all plans and programmes aim at creating optimum employment opportunities for tribal, scheduled class as well as other poor class of people living in and around the forest areas for their socio-economic development. The strategy of the XI plan is to support development of forest resources in the state with focus on the following points:-

- Conserving, protecting and developing the dense forests (crown density more than 40%)
- Regenerating and developing the open forests (crown density 10 to 40%)
- Afforestation and reforestation of the scrub forests (crown density less than 10%).
- Capacity building of the forest department and the village level institutions of the communities to take up protection and management of the forest.
- Promote Ecotourism and Eco development in the protected area. The approach and strategy for tackling the different categories of forest would be different.[10]

EXTINCTION OF FLORA AND FAUNA IS NATURAL:

Biological extinction is a natural phenomenon in geological history. Previously, in every 1000 years, one species was extinct. However from 1600 to 1950 AD the extinction rate had increased to one in every 10 years and now it is one in every year.

This extinction is not due to natural phenomena but due to most dominating species, the human beings.[11] Faunal losses are due to over exploitation of certain species for trading purpose, habitation, alteration and destruction, introduction to new species into the areas and pollution of streams, lakes and coastal zones. This is an alarming situation which needs remedial measures and requires the mentality to understand and preserve all other species as a necessary fact of life. Hence, the need to conserve the vast creation so as to protect all living can survive. We do not have Indian Cheetah now. But Emperor Akbar had 1000 cheetah in his captivity for use in his hunting of over abundant black buck.[12]

This is not a far back, but we do not have any cheetah now. This can happen to any species any where if we will not be conscious enough. Indian Royal Bengal Tiger is now endemic. The Americans have lost their passenger pigeon which once hovered over their sky in millions creating a condition of extensive dark cloud when they flew in the sky. The Indian lion which once roamed from Persia in the west to Kalinga in the east, reported its presence as not too long back as the 19th century in Saharanpur, now languish in a small home territory of a little over 220 square kilometers in Gujarat, yet not very safe of its existence . The Indian Rhino had roamed right from its present home in Assam to the banks of the Indus in seeking refugee to survive. The master male elephants with the herds with their beautiful huge tusk are now a matter of admiration in the museums satisfying man's greed and pride. The future of such master pieces of nature may have to be cultured in the test tubes and cloned using the still available females. There are so many on the brink of disappearance due to different causes.

---

10. Ibid-Page-19
11. Ibid-Page-35
12. Ibid-Page-35

PROTECTION AND DESTRUCTION; BOTH ARE IN THE CONTROL OF MAN:

Due to fanatic approach of naturalists and environmentalists, many species have disappeared. They are trying to save the nature by facing their views on the majority who do not understand their sensitive feelings. The reality as well as fact is that man has come to stay. He is the most selfish, egoistic and arrogant animal. He does not realize that he is nature and everything around him is nature. His numbers of desires are endless and we cannot see any end in sight. He is cascading towards development far from reality and is not feeling the need of any other animate thing in his scheme of development. The fanatics cannot prevent the monsters of progress, which do not realize nor care to realize the future reality of mass annihilation. Therefore there is a need for change in the strategy to save what we can, where we can, rather than believe that if we are adamant to save everything everywhere, we will succeed, but not with arrogant man.

The general trend is that the whim of the nature lover is being bull-dozed with an equal or more accelerating defiance, "What is the need of the hour?" We need a very strong movement of bringing awareness of the dangers that lie ahead and what we would be losing forever. We have to appeal to the hearts for creating empathy for nature. We need to make people realize that why the few fanatics are not opportunists, but are sincere for the survival of humanity and the living earth. We have to realize that if Tajmahal will be destroyed, it can be replicated a hundred times, but not our cheetah, which has gone away forever. Here we need to pay heed to what the governor of Reserve Bank of India Mr. C. Rangarajan, renowned economist had said "Economic growth and environmental preservation are no more opposing objectives"[13] At the "8th Asia-Pacific Parliamentarian Conference on Environment and Development" on 14th November 2000 at Hyderabad, Shri Krishnakant, the Vice President of India recalling the statement of former Prime Minister Indira Gandhi at the First World Summit on Environment at Stockholm 1973 that *"Poverty is the worst polluter"* stated:

---

13. Ibid-Page-36

"the main challenge to day before the global community is not to look at environmental concern and development as an either or paradigm, but as this -as-well-as that paradigm."[14]

The Deccan Chronicle of 31st October 2000 stated that "Every year the farmers in the region (Nizamabad) tend to lose 25% of their yields of paddy, jower, ground nut, maize, cotton and cereals extensively grown, which are both food and cash crops for wild animals." Who will be given priority? The same type of pleas we see, the competition for survival between man and other animals for space and other requirements. We saw the Narmada where a compromise could have been struck rather than a confrontation. Who won? The exploiters, the dominating species on earth that is man we cannot make the brute force majority to realize what they consider as right. We need to resist not challenging but by greater humility as taught by Mahatma Gandhi for the long term success and permanency in realization. We have to educate all who do not know, make them aware and realize their follies. But the environmentalist heard his own voice and saw his names in the newspaper, soon everything is forgotten.

NOT CLANDENSTINE METHOD BUT PEOPLE'S INVOLVEMENT CAN MAKE THE PRESERVATION SUCCESSFUL:

If we think for the preservation for wild gene of tigers, it is more important than the captive born tiger. It may be necessary to sacrifice the latter to meet the needs of those who do not yet realize till they realize. We should be prepared to harvest the captive tigers if need be, as much as we kill without batting of the eyelid, the chicken, ship, fish etc and now the Ostrich, quails, rabbits and crocodiles for meat and skin. Conversationalists need to do rethink rather than be stubborn impractical sentimentalists, weighing the importance of the wild gene conservation by a more pragmatic approach. Our experience shows that when there is a scarcity, natural or artificial created, people resort to clandestine method, but the demand does not declined as long as it can be procured. Instead of decline in demand the price of the

---

14. Ibid-Page-36

product goes up and clandestine activity becomes a lucrative business. It becomes a decimation of the product. Therefore, conservation is successful only when there is people's participation. The best approach is to associate people and create awareness. Association creates interest, interest increases love for nature and love in turn results in desire to protect and preserve. So what better way is there than we love nature, we make others love nature and we all respect and preserve it. There are so many species which are eco friendly and help to clean the nature by eating the dead animals. For example vulture, two or three decades ago they were found in the periphery, but now not even a single. This is not true for vulture, but a good number of species are declining in nature globally. The destruction and fragmentation of habitat, climate change and various anthropogenic pressures, a number of species are under severe stress and whose number is declining alarmingly. In spite of best in situ conservation effort, a number of endangered species are facing the potential threat of extinction.

THE ROLE OF ZOO IN CONSERVATION-BREEDING OF WILD ANIMALS IS BENEFICIAL TO THE ENVIRONMENT:

Zoos, world over, are considered as conservation organizations to serve the need of conservation of vulnerable species, especially where wild life is under severe threat and their survival in nature becomes critical. The responsibility of the zoos for ex-situ conservation of such threatened species assumes greater importance in the above situation. While the importance in the ex-situ conservation cannot be undermined, ex-situ conservation in zoos, aquaria, botanical gardens and germplasm banks can compliment in-situ conservation and serve many other purposes like maintaining viable populations of species threatened in the wild, providing educational and public awareness services, and serving as sites for basic and applied research. More over where the in-situ conservation has no hope for the survival of such species where the threat to wild is now extreme. This method serves many other purposes such as allowing more control over breeding in order to avoid in breeding, increased reproductive rate, providing educational and public awareness programmes and providing materials for basic and applied research. It is an important contribution of the modern zoos to conservation.

Conservation-breeding is a complicated work. Zoo aims to maintain the genetic variety with in a species so that when the species is eventually reintroduced, it is substantially the same as the original wild population. Zoos contribute in many ways to the conservation of biodiversity. There are more than five lakhs mammals, birds, reptiles, and amphibians in the zoos in captivity all over the world. Zoos propagate and reintroduce endangered species. They serve as center for research to improve management of captive and wild populations and they raise public awareness of biotic impoverishment. The contributions of the Indian zoos are significant for the conservation breeding. Nandankanan is the first Indian zoo to successfully initiate conservation breeding programme of Gharial crocodiles. To save them from brink of extinction in its natural habitat, conservation and breeding programme were initiated in Nandankanan zoological park. As no male was available at that moment, a male Gharial was brought from Frankfurt Zoo to initiate conservation breeding programme. It was successful and first batch of hatching was out in the year 1980. They were released in their natural habitat from 1986 in Satkosia and Tikarpara. So also Indian pangolion was successfully breed in the ex-situ of Nandankanan zoo which was very much vulnerable.

Conservation breeding of the vulnerable species is a need based programme. It's breeding in captivity with least human imprinting and release in nature to improve their status in the wild has been a priority in present zoo management. The focus of the ex-situ conservation had been on the bigger animals during initial years. However scientists have gradually raised that each individual species play a vital role through their interaction with others in sustaining the eco-system, which has resulted in expansion of the horizon of ex-situ conservation and all animals are treated equally. Therefore, zoos have tremendous potential for raising awareness about the conservation need of the endangered species of wild life and an understanding of avenues for the recovery of species through changed behavior and philanthropic support.

Excepting conservation breeding of gharial crocodiles and Indian Pangolion, Nandankanan provides natural habitat to different avian species, viz, open bill storks, little cormorant, Night-herons, Egrets etc. with the onset of monsoon, nesting activity attains its peak. The fledglings grow and become adult by October and leave the nest by the end of October. The birds go out in routine foraging to nearby swamps and paddy fields in and around the park. Nandankanan provides food, shelter and safety to these birds for which the heronry is expanding every year. The open billed stork nesting in Nandankanan provided a bewildering experience to its ever growing visitors, naturalists and wild life lovers. But the wet lands outside Nandankanan sanctuary are fast vanishing due to devolvement of new town ships due to expansion of state capital. The abolition of wet lands will destroy the foraging habitat of these birds. Further use of insecticide and pesticide in crop fields around the sanctuary may destroy the snail populations which is the preferred food of open bill storks. Nandankanan is the second largest congregation of open bill stork in the state and first is the 'Bagagahana' of Bhitar Kanika with wild life sanctuary.

CHILIKA; THE HOT SPOT OF BIO-DIVERSITY:

From the stand point of dolphin, migratory birds, Nalabana eco-system of Chilika in our state is considered as the hotspot of biodiversity. The rare, vulnerable and endangered species inhabit the lagoon for at least part of their life style. Nalbana wildlife sanctuary is located within the lagoon spread over an area of 15.53 sq. Kms. It is highly productive eco-system with rich fishery resources. The rich fishing ground sustains the livelihood of more than 2, 00, 000 fisher folk who live in and around the lake. On the basis of rich biodiversity and socioeconomic importance, Govt. of India has designated Chilika as a Ramsar site in 1981, especially as an important water fowl habitat. It is also included in the list of wetlands selected for intensive conservation and management by the Ministry of Environment and Forests, Govt of India.[15] Salient features of Chilika are:-

- Largest brackish water lagoon with estuarine character in Asia.
- Largest wintering ground for migratory birds in India.
- The lagoon comprises the unique assemblage of marine, brackish water and fresh water eco-system.
- A number of rare threatened and endangered species inhabit in it.
- It is the home of Irrawaddy dolphins.

Chilika is considered to be the most suitable habitat for congregation waterfowls in winter and well known as the largest bird assembled sight in the Indian sub continent. Migratory water birds from Arctic Russia, West Asia, Europe, North-East Siberia and Mangolia visit this wet land to spend the winter and to refuel their energy to go back to their breeding ground during the spring. Birds which spend the winter elsewhere in India also use this lagoon as stopover site during their return journey to their breeding ground. In every year, census of birds is carried out by wild life wing for Forest Departments of Odisha in collaboration with Chilika Development Authority in the month of January. Ornithologists and bird lovers from various organizations and educational institutions, researchers, NGOs, local communities and Govt. Officials participate in the programme. The study undertaken by Chilika Development Authority and Bombay Natural History Society between 2001 to 2007 and revealed that over 7,00,000 to 9,50,000 birds utilize the lake annually.

It is most interesting that in the year 2001-2002, 1, 60,000 numbers of Gad walls were recorded in Chilika in one season which is higher than the 100% of its bio geographical population. The bio geographical population of Chilika is 1, 00, 000 in number.[16] Birds in Chilika promote its eco-system in various ways. They help recycling of nutrients by promoting biomass production through growth of macro-phytes. The water fowl in Chilika, i.e; both duck and geese add annually 33.8 tons of nitrogen 10.5 tons of phosphorus through their droppings, which ultimately helps in production of phytoplankton and zoo planktons macrophytes there by promoting production of fish.[17]

---

15. Ibid-Page-48
16. Ibid-Page-49
17. Ibid-Page-48

They also eat the fresh sprouted shoots as well as roots, rhizomes of many aquatic weeds in the lagoon. Another special attraction of Chilika is Irrawaddy dolphin. One can see it in its natural habitat round the year. In 2000 these were seen in the four places of ckilika lagoon, i.e; Kalijai, Nalaban, Pathara and Satpada. But now they are seen in various parts of Chilika. They exhibit seasonal migration between various ecological sectors of Chilika lagoon. As per the annual census of Irrawaddy dolphin in Chilika during February 2008, their number is 138 and 30 individuals are in Bramhani and Baitarani of Bhitarakanika. In India, they are protected under wild life protection act. The major causes of their mortality are entanglement in the gillnet and other fishing gears, injuries from collision of motorized boats, sharks bites, neonate deaths and diseases.

Therefore, training and sensitization programmes are conducted for the awareness of the boat operators as well as to follow the dolphin watching protocol developed by Chilika Development Authority. Propeller guards are provided to boat operators for fixing in the motorized boats free of cost to minimize death/injury to dolphins. Awareness programmes have also been launched to sensitize the fisherman as well as the visitors for protection and conservation of dolphins in Chilika. Signages are placed in strategic locations creating identified for conservation and management through regular monitoring on monthly basis. Scientific studies have been initiated in anatomy, habits and ecological characteristics of dolphins. Studies on underwater behavior of dolphins have been taken up in collaboration with URA Lab of University of Tokyo and IIT Delhi through deployment of hydrophones. Genetic study on them has been also initiated in collaboration with central Institute of Freshwater Aquaculture, BBSR.[18]

However Chilika provides shelter to various flying and aquatic animals, thereby proves as the spot of biodiversity in the map of our country. We should be cautious enough to keep it up for the future generation as well as the name and fame of our state and the country. The artificial prawn farming developed in the lagoon due to its

---

18. Ibid-Page-50

high market price is polluting the area by affecting economy of poor fishermen, those who are depending on it from generations together. Govt. is making policy; people are gradually becoming conscious in order to restore its lost environment.

Since policy is very broad based and unsupported by law, it is the administrative orders which become main planks for discussing the improvement in policy. So the people should be involved for the protection as well as management. Mutual productive relationship addressing legal and social dimensions has to be evolved. Many day-to-day operational issues are likely to arise in such type of programme. There is a need to segregate issues for different level of decision making. Members from NGOs, beneficiaries should be included in making decisions. It should be maintained like adjacent forest areas with active participation of local people.

THE CONVERSION OF DEGRADED LANDS INTO FORESTS FOR ECOLOGICAL BALANCE:

Apart from forests and lakes, there are various types of degraded lands like Govt waste lands, lands own by communities, privately owned degraded lands etc. in our country. Local yields from degraded forests and greater demand are not compatible.

Too little benefit in too distant future does not attract poor people to forgo their immediate needs in the hope of future uncertain rewards. As they are very poor, their needs are largely with immediately survival concern. It is evident from the fact that people participation has become sustainable whenever subsistence need material like grass or non timber forest products have benefitted poor with low period. In view of this, there is a need to develop an integrated programme around all degraded land resources including social or farm forestry through extending forest frontiers outside the forests.

One of the main reasons which have degraded forests is higher removal from forests than its annual increment. Main reason is that a large number of communities, some of them unauthorized take away materials under the pretext of rights and privileges. This has made forest an open-ended resource. At the time of settlement, forests are assigned to specific villages and only communities from such villages have rights and privileges. Forests are demarcated also in this manner on the ground. But this fact has been lost sight of reducing benefit to communities. Privileges without corresponding responsibility are counterproductive. However Joint Forest Management Committee is offering unique opportunity indirectly making forests close ended resource. Their area of operation should be demarcated on the ground to ward off future dispute of ownership. And the village level user groups can transform forests into a closed ended resource, which would bring more benefit to local user group.

The rural population depends on forest for fuel wood, fodder, small timber and non-timber forest products. The fuel wood was the main source of energy in rural India. It's demand increased with rapid increase in population. The estimated demand for fuel wood in 1976, 1996 and 2006 was about 133,200 and 240 million tons respectively (Planning Commission 1985). The maintenance of cattle, sheep and goats of forest dwellers and nearly forest dwellers depend on forests largely. The fodder is a free supply and the forests are being over exploited. The rural people as well as the forest dependent communities depend on forests for small timber for construction of their houses. It was estimated that the use of timber for construction of houses, agriculture, furniture and packaging about 54.4, 60.4 and 66.6 million m^3 in 1996, 2001 and 2006 respectively.[19] The demand is fulfilled partly by the forest and the rest from the trees growing outside forest. Likewise bamboo is also used for house and households. It is estimated that 16 lakhs tones of bamboo are used every year for that purpose. State government also supplies bamboo to Artisans and Co-operative Societies at concessional rate. The govt. provides certain rights and concessions to the people for use of forest products. Let us have discussion on it.

RIGHTS AND CONCESSIONS FOR USE OF FOREST PRODUCTS:

After independence, expectations of people to get more and more from forest land increased. To win the election, the rights and concessions were liberalized which induces people for arbitrary and irrational use of forests resources. The liberal concession in one integrating unit became the norm for other units, the discipline of

---

19. Ibid-Page-61

forest management slackened and the demand on forest resources intensified. Govt. was unaware of the serious consequences that rose due to exercise of these rights and concessions. But after 1950 many states set up high level committees to go into the complex problems, as a result very valuable reports emerged. The political will needed to implement the recommendations was however lacking. While the Govt. was granting such privileges to the public, the carrying capacity of the forests was not taken into account. So the forests in the neighborhood of the villages had disappeared or badly affected and were unable to bear the burden of the required quantum of rights and privileges or that the population had increased was not taken into consideration. Therefore, all attempts for conservation of forests were undermined.

The inflation, that the country faced in the seventies led to abnormal increase in prices of timber and fuel wood. The forests which were fast dwindling and were unable to meet fully the joint demands of commercial interests and villages, came under the axe of timber smugglers as well as ignorant tribals, who were tools in the hands of unscrupulous traders and middle men. With the declining supply of timber and firewood, prices sky rocketed and exploiters multiplied in a geometric progression.

After independence, the country faced serious shortage of food supply. The increase in food production was sought through expansion of area under agriculture at the expense of forest areas. A climate was created that forest lands must be used as agricultural purpose. Encroachments that were started in this atmosphere were fanned by political groups and vast areas under natural growth were cleared. Govt.'s policy to settle encroachment encouraged more and more people to cut forests and grow crops. In some districts, the forest areas dwindled to one fourth, as a result of these encroachments and settlements.

The rights and concessions to collect and remove fallen and dead fuel wood by head load from forests for bonafide use was available almost throughout the country. In some cases a nominal fee was collected for it. Such type of privilege was available to neighborhood villages of forest. But gradually it was slackened. The head loads were permitted to other villagers, urban people. Except bonafide use, sale and resale was also permitted. Green trees were also converted into fuel wood and these were exploited by head loaders and agency of head loaders. There were passenger trains connecting stations passing through forest areas full of fuel wood head loads. Similarly grazing became an uncontrollable menace. Either the grazing units disappeared or the enforcement of regulations became difficult and cattle grazing all over the forests, irrespective of their carrying capacity became the normal practice.

On the other hand, the obligation to the forest department for protection of forest from forest fire etc. was gradually forgotten and the enjoyment of rights and concessions not only continued but multiplied. As the forest degraded, the department of forests started stricter measures for the conservation of forests. The pace of forest degradation increased and by1970, about one third of the forests were suffering from degradation. It became clear that forest conservation was not possible without the involvement of people. In 1998 Natural Forest Policy was framed, where specific provision was made for public involvement for protection and management of forests, which became successful.[20]

ENACTMENT OF LAWS FOR SUBSISTENCE OF FOREST DWELLERS AND PROTECTION OF FOREST:

Indian forests are rich with forest products. They produce Kendu leaf, Mahul flowers, Sal seed, Harida, Bahada and Anala and fruit, root, tubers, honey, meat of the wild animals etc. The tribal people chiefly depend on forest for their livelihood. The people who live around the forest chiefly depend on forest for minor forest produce. It is estimated that 60% of minor forest produce are collected for household consumption as well as income generation. Collection of these items gives employment to all men, women and children at their door steps.

National Commission on Agriculture estimated in 1976 that the employment potential to be more than 25 crore man days. So, minor forest products play greater importance for forest people.

---

20. Ibid-Page-63

An act was passed in the Parliament in the name "The Scheduled Tribes and other Traditional forest dwellers (Recognition of Forest Rights) act 2006 No-2 of 2007 to recognize the forest rights and occupation in forest land in forest dwelling scheduled tribes and other traditional dwellers who are residing in such forests for generations, but their rights cannot be recorded. The act came as the tribals and forest dwellers those who were living in the forest land their identity were not recognized which was a great injustice to the forest dwellers. The act not only provides right, but includes responsibilities and authority for sustainable use, conservation of bio-diversity and maintenance of ecological balance and thereby strengthening the conservation regime of the forests while ensuring them food security and livelihood.[21] The preamble of the act says so. Gram Sabhas are to be framed to constitute committees for the following activities:-

- The protection of wildlife, forests and bio-diversity from among the members.

- The protection of water sources, catchment area and the ecological sensitive areas.
- To protect the cultural and natural heritage of scheduled tribes and other forest dwellers.
- The implementation of decisions taken in Gram Sabha prohibits any activities which affect the wild animals, forest and bio-diversity.

LIVELIHOOD ISSUES VRS. ENVIRONMENTAL CONCERNS:

Forest plays a pivotal role keeping the economic stability of forest dwellers as well as forest neighboring villagers. But the ever-increasing human population for the search of livelihood opportunities has surpassed the carrying capacity of the fragile eco-system. Human greed of large population is threatening to make the system unsustainable. So long as the nature has the capacity to tolerate the human perturbation, it could.

---

21. Ibid-Page-64

But in the recent years, the adverse impact has been bringing about irreversible changes as per the doctrine of "Survival of the fittest", the more vulnerable species, both flora and fauna are disappearing or in the brink of extinction, which would collapse the eco-system sooner. Here two issues come to our mind, which appear to be conflicting in nature. One is Environmental and another is Livelihood. The conflict arises when the methodology of harnessing livelihood affects the environment adversely.

There are stake holders who involve in the "Environment versus Livelihood Issues are of three types.

a) The advocacy group
b) The Community
c) The Government

These stake holders have great influence in formulation and implementation of strategies.

a) **The advocacy group**

The advocacy group is divided into two groups. Some who pleads for total non interference in the eco-system. The other group pleads for giving the livelihood opportunity priority over everything else. These two groups are again divided into three groups; viz, good, bad and ugly. Those who understand the problems and are committed to the cause with positive attitude are called 'Good'.

Those who have limited knowledge on the issues but obsessed about the issues basically with a negative mindset are called 'Bad'. Third one is vested interest group, which have their hidden agenda and have their own axes. The people under good group understand the problem of eco-system and threats upon it, then find the positive solution. The people under bad group are guided more by heart than by head and far from actual problems. But the people under ugly group are the most treacherous fellows. They have the correct understanding of problems but disclosed with vested interest by giving certain colour. By way of charismatic influence, they manipulate the government machinery, influence the media and at last become successful in opinion building to achieve their long term disguise objectives. If at all their mischief is exposed, it has become too late and the issues are no longer under the scanner.

b) **The Community**

Different types of people live in a community. It has limited access to information resources. The community may be sub divided in the following way.

- The set of people whose primary livelihood opportunities are at stake.
- The faction that is dependent on livelihood opportunities.
- The people who get used by the various vested interest groups both from within and outside the community.
- Local opinion makers, who may or may not be having in direct interest in the livelihood issues. Each group has its own perception and understanding of the situation. So all the sub groups should not be mixed into one category while finding solution to the problem.

c) **The Government**

The government has the responsibility for both conservation of the environment and providing livelihood opportunities to the people. In case of any conflict, it has to formulate the best possible strategy and implements the strategy. As legislation and execution are two parameters of good government, it must first spell out its policy without any ambiguity and then execute it in both letter and spirit. But it must be kept in mind that there are tentacles of vested interest groups who are well established in the government and they have the potentiality of influencing both legislation and execution. These parasites are so dangerous that they cannot be easily traced out and driven out from the machinery. Their cross cultural activities like saying in same way and doing in different way brings enormous harms to the country. All the plans and programs of the government become non-productive. So if the vested interest groups would be identified and kept aside from government machinery while preparing policies, executing and implementing policies then the whole country would reap its good result and more benefit would be achieved by the government machinery for the public.

FOREST AND ENERGY:

In ancient times people were using wood as energy producing element. But after discovery of fossil fuels and industrial revolution, it became the major source of producing energy and now-a-days it is at the threatening situation. In India 80% of fuel energy, i.e.; fossil fuel is imported from outside. The rate is determined by them and as it is used in the vehicles, so also the price of the commodities are determined accordingly, which affects the economy of our country. So many countries are trying to use bio fuels which can be regenerated and produced inside the country by using the degraded lands and wastelands. If technological development makes it more efficient, at least economical to produce liquid bio fuels from cellulosic material, the result would be an increased in energy efficiency and improved overall energy balance. At present wood energy is most competitive when produced as a byproduct of the wood processing industry. Wood residue from felling and processing operations generally constitute more than half of the total biomass removal from forests. In natural forests, up to 70% of total volume may be available for energy generation. Where human and financial resources are limited, bio energy takes the first opportunity of development. If sufficient crop land is available to produce food at affordable price and to avoid loss of valuable habitat, it is clear that bio energy strategies are closely linked with and integrated in agriculture, forestry, poverty reduction and rural development. It is proven that carbon emission and green house gas emission is less than petroleum fuels in bio fuels. So the present situation demands from the forestry sector a new dimension of energy supply, mitigating climate change and supporting sustainable economic and environmental development.

In order to preserve ecology of nature, forest and people are dependent on each other and ought to grow together. We should participate actively to combat all environmental hazards to avail its eternal beauty and grandeur. But growth of environment without forest is inconceivable and also tends to be negative. Since the time immemorial, wood is practically the most practical solid fuel for heating and cooking requirements. The rate of urbanization has increased the rate of fossil fuels, gas, electricity in urban sector, but the pitiable scenario still persists in rural bases. More than 70% of total population resides in rural areas and their full requirement is primarily met from tree growth, though some of them use agricultural waste, cattle dung. Even if, developed countries, the magnitude of usage of wood energy cannot be zero. Recent statistics say that in USA, 2% of total requirement of energy is met from wood. In Sweden 8%, 15% in Finland and 27% in Brazil etc.[22] Apart from these, wood is used for manufacturing ply board, chip board and paper etc. Wood can also be converted into various gases and liquid product particularly methanol for certain engine requiring high octane fuels. Methanol can substitute the motor spirit during the time of shortage in transport sector as well as agricultural machinery. It can also check and control the sky rocketing crude oil import bill by using the renewable source of energy, i.e.; wood, which have equally contributed for rural development and national growth. Therefore, it is high time for us to concentrate with wood cultivation at par with agriculture. We can also mobilize man power in rural sector in increasing national income. Though attempts are made for Agro Forestry, Social Forestry, Farm Forestry, Avenue Plantation, Block Plantation, Man groove Plantation etc. and these are encouraging and successful also, still continuous and perpetual effort should be made to restore the beauty of nature and protection of environment.

The secondary forest fuels cater to the energy needs of the industrial sector and can save a lot of fossil fuels. They also create greater employment opportunities with very low capital and skills required for their management and found in the local community. If emphasis would lay on wood energy, then it will ensure greater social justice and equal distribution of national wealth, which will help for elimination of poverty and unemployment in our country. The tropical countries have enough reserves of such renewable energy due to heavy sun light. It should be methodically exploited to meet the consumption needs which would also leave surpluses in the form of charcoal, briquettes etc. So from the holistic and rational point of view forests are the major store

---

22. Ibid-Page-74

and supplier of energy. So Bio-Energy is only dependable, safety and endless energy before us as compared to other energy. The wood energy mainly benefits the weaker section of the society trough mass employment. There is no dearth of technical personnel, land and water, sun light and human resource in our country. Public awareness and political motivation are necessary to materialise such projects. Oil and fossil fuels are finite source of energy. These will be finished within four to six decades where as wood is ever-so-ever, a renewable source of energy. So let us start, tomorrow is too late. Forest maintains the equilibrium in Eco-system. The change of climate is always threatening the life forms of earth.

According to survey made by Forest Research Institute, Dehradun, forest gives us various benefits including timber, fuel wood, bamboo and medicinal plants.[23] These are as follows:-

    a) Production of Oxygen   15.20%
    b) Control of soil erosion and fertility conservation   19.90%
    c) Conservation of animal protection   02.30%

d)	Help in recycling of water, improving humidity	19.60%
e)	Protecting birds and animals	13.00%
f)	Control of air pollution	30.00%

                                                    Total 100%

Various reasons like population explosion in our country, excessive biotic interference, illicit felling of trees, mining activities, hydrothermal projects, urbanization etc have brought damage to forest resources. Bio diversity is necessary not only for human beings but also for scientific documentation of flora and fauna. In odisha, tribals have destroyed vast areas of natural forest in the western part and only barren hillocks are found due to regular illicit felling of trees. The wild life in the forest is going down due to damage of forest and poaching, for which visitors fail to see animals in the forest at present. But now some sorts of consciousness have been created among people for protection and renovation of forests.

---

23. Ibid-Page-76

The coastal jungles, which were destroyed by people, are also renewed by the government of our state by restoring management on the forest department by making different divisions and ranges. The only objective was protection of forests under all adverse conditions. Therefore, following measures should be taken for protection for forest as well as wild life.

- Effective mobile squad with armed staff should be formed and operated.
- Theft prone area should be identified.
- Foresters and Range Officers must be provided modern vehicle facility and VHF.
- For collection of information, intelligence wing should be formed under the control of DFO.
- Frequent patrolling should be made on group wise.
- Theft prone areas should be inspected by DFO frequently.
- Village committee should be formed and conduct meeting regularly.

- Assessment of the damaged forests and its development should be done by forest officers.
- Forests should be protected from fire.
- Immediate action should be taken against the culprit for illicit felling and poaching.
- Eco-tourism system to be adopted in Sanctuary and National Park area, which will create employment for local youth and also control illicit felling and poaching.
- District Collector and SP should review the problems and take steps for protection of forests.
- Political patronage should be provided to both govt. and non-govt. persons, those who work for the safety and security of the forest.
- People should be honoured with prizes or awards, those who work for the safety of the forest or have jeopardized their lives fighting with offenders of the forest.

MORAL APPROACH:

Thus far, we have highlighted the necessity of government measures for the protection of environment and ecology. But the striking features which we see in every sphere of constructive change are moral bindings that the man in the society must possess. If man has moral bindings in his mind for the commitment of some issue, then that issue will be immediately solved.

Taking in view, the above mentioned fact, some thinkers in our society, have offered life to inanimate objects, so that we would be able to protect, as matter of sympathy, to our surroundings. Take for example the case of Niyamagiri of western Odisha, court ordered to arrange Gram Sabha, for the permission to collect Bauxite by the Vedanta Company. In the Gram Sabha, each member demanded Niyamagiri to be his God who has a broader life and who nurture the inhabitants of the area. As a result, in the Gram Sabhas, a moral binding was found in every individual, which resulted in unanimous decision to protect Niyamagiri from plunder and exploitation.

Just like that a saint named Bhakti Keval Audulomi said:-

Shri Gouda mandala bhumi yeba jane chintamani,

Tar haye braja bhume basa.[24]

It means he who knows that the soil of Nawadeep is also conscious; he is eligible to reside in Brajabhumi, which is Golak Brindaban. Here also the inanimate soil is given life so that man will have moral binding to be detached from exploiting the nature.

In the Puranic age, in our scriptures and religious books, we see several instances to offer life to the soil, to the stone, to the trees to the sky etc. So that a moral culture to protect environment and ecology will develop.

The above mentioned explanation is highly moral and philosophical for the preservation of our surroundings, ultimately creating a culture to protect environment and ecology. The matter does not remain in whether the stone has life or not, but the matter remains in whether we can protect the mountain or not. It is for man to decide what to do and not to do.

---

24. Dham-Parikrama

****

CHAPTER-V

COMPARATIVE STUDY OF CHAPTER III & IV

COMPARATIVE STUDY OF CHAPTER III & IV

RATIONAL USE OF NATURE & NATURAL RESOURCES TO SAVE THE EARTH FROM PERIL:

If human beings will become rational in every sphere, environment will never further deteriorate, rather be renovated. Though it is a troublesome work, but we have to take it. The developed should think about underdeveloped, rich about the poor and haves about the haves not. We are living in one and same planet. We have only divided into different parts; as countries, states etc. from our point of view. When some parts of our body are affected, the whole body takes the pain. If America will emit more carbon gases, the whole world will suffer. What we are calling global warming, is the result arising out of it. Nature is uniform and harmonious. It does not discriminate between human/nonhuman, good/bad, rich/poor etc. Therefore, there must be uniformity in our saying and doing. The difference between these two gives rise to cross thinking. Different conferences are convened to save the earth by controlling carbon emission. But those countries, who take the leading part are practically not agree to leave their interest. They are crying for protection of environment but doing which will bring harm to it. So doing mechanically and thinking philosophically put them on the cross road. Therefore, there should be coherence between action and thought for the benefit of the environment. So there is a saying; it is better late than never. Though it is not done, but we have to take it now.

Human beings have exploited the nature and natural resources without thinking about the future consequences. The unmanaged and gigantic uses of water, soil, minerals and forests have direct effect on the climatic condition of that area. Only within few decades, biotic pressure has increased on earth. During past hundred years, human population grew in billions, i.e., 1.6 to 6.5 billion and cattle population increased fourfold than human population resulting adverse effects on natural resources like depletion of forest or so.

But now time has come, we all should make sincere efforts to restore the loss and hand over the living planet to our future generation in green, clean and serene condition. The mentality of our children should be turned, now, to protect, preserve and promote natural resources for posterity. We should begin our days with a thought of greening surroundings and ends up in conserving nature. It is perceived that the excessive use of fossil fuel is the cause of global warming leading to change of climatic condition. We experience that during last two decades, there has been gradual increase of atmospheric temperature, i.e.; very harsh summer months and mild winter period coupled with erratic and accidental rain fall during the year. The global warming is due to green-house-gas-effect, which includes carbon dioxide, methane, nitrous oxide etc. Our mother earth is covered with 75% Nitrogen, 21% Oxygen and balance in the form of green house gases. Abnormal presence of green house gases cause air pollution in the atmosphere like a thickening blanket and traps the heat of the sun. The rise of temperature was 0.74 degree Celsius by the end of twentieth century, which caused the rise of sea level, melting of glaciers destroying species and producing extreme weather.

The sources are well known to each and everybody, i.e.; burning of fossil fuels in different auto mobiles and industries whose emission is harmful gases, destruction of forests, burning of organic matters etc. These activities contribute to the increase of green house gases. The decomposition of organic matters in agricultural practices, dead bodies, dung etc. release methane gas. Chemical fertilizers applied in agricultural fields, produce nitrous oxides which are harmful to the atmosphere. The developed countries like America, Japan, European nations and developing countries like India, China, Brazil, Russia release maximum green house gases. USA is the largest green house emitter which releases about 6.9 billion tons of carbon gases per year. Data reveals that during the last 150 years (1850 to 2000) out of total CO_2 emission in the world, America emits 30%, European Union 27%, China 73%, but India 2% only.

Scientists of all over the world predict various consequences affecting our planet due to global warming. The mountain peaks covered with ice and polar ice will be melted; leading to rise in the sea level, which will result in coastal flooding and the major cities of the world situated on the coastal belt will be submerged. It will also lead to shifting of habitation due to submergence and new emergence of migrant people at other places will create social disturbances by affecting their economy & culture. Different natural disasters like Tsunami, Hurricanes will take part due to warming of sea level. Rivers will be dried up very soon just after rainy/winter season, which will affect agricultural production, loss of fish and aquatic eco-system. Global warming has extinct many animals and plants from the planet every year.

DIFFERENT MEASURES TO CONFRONT GLOBAL WARMING:

Irrespective of single-joint, rich-poor, literate-illiterate, country-society; effort should be made to check and control global warming. There need to be changed in the mind-set of policy builders, businessmen and industrialists to feature the environment in growth policies and economic success. Few pragmatic steps to be followed by one and all as such:

- Modern/ Green technology should be developed through scientific research to cut down carbon emission by industries.
- Deforestation should be controlled, forest wealth should be conserved, new forests should be created in the extinct areas as trees act as carbon sink by absorbing CO_2 from atmosphere during photosynthesis and thereby act as lungs of the country.
- Natural resources like water and mining materials should not be wasted, but properly used.
- Cost effective and eco-friendly vehicles should be used by replacing the old vehicles, which are emitting smoke and creating rough sound, which creates both air and noise pollution.

- More emphasis should be given on renewable energy such as Sun, Wind and Water. To discourage use of chemical fertilizers, insecticides and pesticides in agricultural practices and practise organic farming.
- To use CFL bulbs by replacing incandescent light bulbs which are energy saving and environment friendly.
- Rules and regulations should be strictly followed by the government against polluting industries.
- Coal burning power plants should be replaced by cleaner plants. Emphasis should be given on hydro electricity and nuclear power plants for generation of energy. But after Fukushima nuclear accident at Japan, the people have turned their faces from nuclear power to other sources of power generation system.

Visualizing the environmental degradation, world leaders have decided to meet frequently to save the planet from different disasters, such as:

KYOTO PROTOCOL

Visualizing the dreadful effects of green house gases, United Nation convened a conference in Kyoto, Japan from 1st to 22nd Dec in 1997 to control industrialized nations through legal binding regarding emission of green house gases. The protocol was opened for signature on 16th March 1998. It's main aim is to curb carbon emission by at least 5.2% below 1990 levels in time frame of 2005 and establishes three mechanism for reduction of green house gases like (i) Clean Development Mechanism (ii) International Emission Trading (iii) Joint Implementation.

COPENHAGEN MEET

In Copenhagen, a meeting was held from 7th to 18th Dec, 2009 regarding change of world climate. Probably, World is the only planet; where life is possible. Due to increase of climatic temperature, the small Islands and coastal places will be drowned. So they started to create consciousness among the people of the world in various ways. The President of Maldives convened a meeting of his ministry in deep sea from 35 nautical miles from the coast and it was continued up to 45 minutes. The member of his ministry attended the meeting using oxygen mask and requested the people of the world to control global warming. If the water level will rise from 18 to 20 cm, this country having on area of 800 sq. kms of 100 Islands will be drowned in the water. So also the govt. of Nepal organized a meeting at the height of 5250 mts. in Himalayas to create consciousness just before one week of commencement of Copenhagen meet. They worship it as the king of mountains. 100 crores people of Asia depend on it for their maintenance. Due to global warming, the ice stored since thousands of years will be melted and the rivers like Ganga, Bramhaputra, Jamuna flown from it, will be dry after rainy season. So the people of Nepal, Bhutan, Tibet, China, India and Pakistan will be affected by it. In this context, they published "Everest Declaration" before Copenhagen meet.

In Copenhagen, the representatives of 192 countries, 110 heads of Govt., thousands of scientists, officers, experts, advisors and volunteers attended. The aim was to control the developed countries from emission of carbon di-oxide, transfer the advance technology to developing countries for lessening of carbon emission and persuade those countries who have not signed the Kyoto protocol. Those countries, which have taken steps for reduction of carbon and created carbon sink should be given financial incentives. Before the commencement of the meeting, a document was prepared by Denmark, Australia and America under the leadership of America called as "Framework for Action", where they have reduced 30% carbon emission to 3% abolished the legal obligation and made according to the necessity of their country which was antagonistic to the pervious decision. But they could not produce it in the meeting, because it was disclosed in the English Daily News Paper '*Guardian*'. African countries wanted to boycott the meeting, because they wanted that discussion should proceed on the basis of Kyoto Protocol. When Mr. Obama, the President of America saw this, he wanted to prepare an agreement; otherwise his principle of public relation will be futile. Some countries like Brazil, South Africa, India and China were consulted in closed door by Mr. Obama and prepared an agreement, after those other 25 countries were consulted. Though, Copenhagen meet was a great challenge to control climatic changes, but resulted in vain. The developed countries wanted to minimize the increase the world temperature within 2^0 centigrade (small countries 1.5^0) with the advice of scientists having no such policy how to control it. No bar for carbon emission was there for industrially developed countries or any legal binding. In Copenhagen meet, the emission of green house gas was totally kept aside. The difference between haves and haves not countries was also wiped out. The developing countries like Brazil, China and India agreed to reduce the green house gases from 20 to 25% by 2020 voluntarily, but the developed countries were requested to declare their rate of emission by 31st Jan, 2010. It was decided in the meeting that, developed countries should provide 130 billion dollar to developing countries for development of green technology and out of it, 30 billion to be provided in 2010 with a condition that, they should sign on the Copenhagen agreement. But a small Pacific Island country 'True Value' protested against it by saying that "we are compelled to sell our future in exchange of thirty pieces of silver". Venjuella told that, "Our policy is not made for sale. You keep your cheque book in your pocket and tell us clearly what the policy of carbon emission is?"

According to great environmentalist Sunita Narayanan, this agreement cannot be told as agreement on climatic change, rather it is an agreement made to justify the right of pollution. Though Copenhagen meet was not fully successful, therefore all waited for Cancun meet to be held in 2010.[1]

CANCUN SUMMIT

"Cancun summit was held in 2010 according to protocol, but ended with no success. All the world leaders delivered long speech with crocodile cry. But rich countries did not show any sympathy for the protection of earth. The people of rich countries are so intoxicated with luxury that, they are not interested to hear the problems of poor, those who are climbing the stage of development, because they have reached at the top level of the progress. No clear picture was given for creation of fund for transmission of green technology to poor countries. As Copenhagen meet was unsuccessful, so everybody had an expectation from Cancun summit; but in vain. Rather it was shifted to Durban meet of 2011which hurt the environmentalists very much. India declared from its own to obey all the decisions taken legally. NASA has declared that

2010 is the hottest year among 131 years. So the ice is melting alarmingly in the North Pole. Scientists say that, if it will continue, the polar zone will be iceless within 20 to 30 years during summer. It can be best thought, what would be the situation of ocean. So in this juncture, our leaders are not caring for the protection of earth.

---

1. Odia Daily News Paper "Dharitri" dt.14.12.2010.

We have to remember that even if environment will be destroyed but the earth does not. Neither we nor other flora and fauna would be there to make this earth beautiful. Therefore, today we have to decide what type of earth we want to see in future. All our activities clearing the way to convert this beautiful living earth to dead planet."[2]

So govt. of India is taking different steps inside the country for the protection of environment, such as National Action Plan of India. However, Prime Minister of India Dr. Manmohan Singh has launched National Action Plan on climate and the key elements are as follows:

- By 12th plan, solar energy to be boosted by 1000 MW.
- Through energy efficiency management, 10,000 MW to be saved by 2012.
- Industries like steel, textiles and power are expected to trade in energy efficiency targets.
- Recycling of automobiles at the end of their life.
- Coal thermal plants to be closed gradually.

Apart from it, Govt. of India, Urban Development Ministry has taken different steps to avail carbon credits, i.e., energy efficient building, mass rapid transportation, solid waste management and sewage treatment etc. Himachal Pradesh has mandated to begin environmental audit and on its way to become the country's first world's carbon-free State.

Global warming is the current issue which is irritating the people world over. So both developing and developed countries are trying to do best of their efforts as to how progressively reduce emission of green house gases. So need of the time is, all of us should be eco-friendly in our life style and use non renewable materials sustainably, for our own sustenance as well as our successors. It is our moral duty to be conscious and make others conscious about it.

2. Odia Daily News Paper "The Samaj" dt.18.12.2010

ECONOMIC DEVELOPMENT MUST COINCIDE WITH ENVIRONMENT:

It is a matter of great regret that our political leaders are inviting the foreigners to establish industries like steel, powers, aluminum etc. by using raw materials from our state. Further, some of them are given the facility to establish special economic zone (SEZ) facility and Port by which they will be able to exploit as they can. Though, there are some checking facilities, that is only nominal. They can tape the officials through different means and will be able to be success in their mission. Now, different metals like, coal, iron, bauxite, manganese are lavishly stolen by the mafias with the knowledge of some corrupt officials and political leaders and exported to different foreign countries like China, Japan etc. The govt. is silent as if does not know anything. But when the conscious people are raising their voices through different Medias, govt. becomes alert and finishes its duty by establishing commission and suspending some officers in order to show it's cleanliness before the innocent public. But the white elephants who are taking lion's share from that business, remained left. The opposition parties who shout inside and outside the Legislative Assembly, they also become silent. The cause is well known to everybody. The only thing is that, our state is losing its revenue and materials which went away to other countries, cannot be formed newly, which could be used by our successors in future.

Here, I want to mention that most of the iron ores were exported to China and Japan. China raised the rate of ores and took at the highest rate. Japan has no place to store them on land. Therefore, it stored the iron ore in the sea. It reveals that how conscious they are about the future of their people. Another thing is that, our govt. is inviting multinational companies to settle industries in the name of economic development providing valuable land, water, raw materials by ousting our own people, polluting our clean environment. Our state govt. as well as central govt. are showing much care about them. Only local people, those who are expected to be affected are shouting and opposing them. Govt. is using police force against our own people, those who are opposing the project work. The argument is that, govt. is working for the benefit of the state.

Again, we all know that the benefit of the state means the benefit of the people. Then the question arises, if the project is for the people, why are they opposing? The answer is that, the opposition is not by all, but by a few people of that locality. They should be invited into discussion, after consultation with all and considering all the dimensions, the project is to be undertaken. In exchange of valuable land, raw materials and environmental degradation, if our people will not get economical benefit, why shall the project is permitted? There should be cleanliness in every sphere while undertaking a mega project. If the corrupt leaders by taking bribe from companies permit for establishment of industries by ignoring the interest of locality, state etc. then there will be serious agitation in spite of govt. approval. Now the case is going on with POSCO and Vedanta in Odisha. People are anticipating that they will lose domicile land, livelihood, fertile agricultural fields, in terms of minimum compensation and nominal job. In case of Vedanta Alumina project in Lanjigarh of Kalahandi district, the indigenous people those who will be ousted for the project are opposing seriously with their traditional weapons in anticipation of loss of their livelihood, environment etc. Different organizations as well as judiciary supported their demand and state govt. was harassed. In case of Vedanta Viswavidyalaya, govt. of Odisha signed MOU with Vedanta Company and started acquiring land at Puri district, but could not succeed. It is a matter of regret that govt. had transformed some agricultural lands of Lord Shri Jagannath which was refunded after strong protest from all the sides, which reveals the evil intension of govt. In Kalinga Nagar of Jajpur district twelve indigenous people were killed by police force while acquiring land for TATA Company. The matter was seriously viewed all over the country.

Such type of case happened at Singur and Nandigram of West Bengal, while state govt. was acquiring land for Tata Company. The local people protested seriously, the govt. refunded the land to the people. It is said that, the companies, which are coming to produce iron from iron ore will not be helpful for the economic development of our people. If they will produce other engineering goods from the product then that will be helpful for our economic growth. China produces 400 million tons of iron per year. But other engineering goods produced out of it are valuable, which add to the economic benefit of the country. If the use of water, mining materials, had adverse effects on the environment, it will not be taken into consideration. These iron/aluminum industries are not benefitted for us. So the companies should be compelled to set up assisting industries for production of engineering items by using the products like Alumina or iron. There should be cleanliness, sincerity and sense of future responsibility to the public under adverse situation, before signing MOU with a company. Now-a-days the people of Odisha consume 3500 MW electricity per year, but govt. of Odisha has signed MOU for production of 15000 MW per year. We all know that, most of the electricity is produced out of thermal power stations. Again 97% of power that we use in different ways for our daily use is received from burning of fossil fuels. If more electricity will be produced, more coal will be burnt. In the present situation, the temperature of our state is increasing alarmingly during summer. People are suffering a lot and dying out of sun stroke etc. When more thermal power plants will be established, our situation will be graver. Is it not proving the in-differentness, irresponsibility of the govt. about the environmental degradation? When people are suffering a lot due to environmental hazard; learned persons as well as experts are warning through different media, govt. is not paying any heed to them rather doing according to its sweet will. If people will suffer, die, then who will enjoy the result of economic development. What is its meaning? In the name of economic development, we cannot destroy our environment. If environment will be affected adversely, we have to bear its result. If more and more industries will be set up, a few people will be benefitted, but thousands will suffer. The people of that locality, where industries are set up will suffer from various diseases due to dust, smoke, effluents etc. Not only people, so many non-human creatures like birds, animals will be affected or extinct. Generally industries are set up in remote areas or forest areas. When forests will be depleted, different wild animals will die by entering into village areas etc. Habitats of forest animals will be destroyed. Indigenous people or forest dwellers will lose their livelihood. If so, many things will be disturbed for benefit of a few people, how can we say that it is development? Are they not part of nature? Who gave us superiority to disturb the equilibrium of the nature? We should not think ourselves as master of nature, rather a member of nature. Such type of thinking will disorder the peace, tranquility and harmony in nature. We are the creation under the same creator. Again, if we are getting pleasure by visualizing the diversity of nature, it is our duty to leave in the same condition for our future descendents. If we cannot add something new to the nature, we have no right to subtract from it. Therefore, our duty is to do good and be good with nature. It will be possible only when there will be harmony between action and thought. [3]

DEVELOPMENT SHOULD BE HOLISTIC AND VIRTUOUS:

The vice-chairman of Planning Commission of India Mr. Mantek Singh Aluwalia had once said that, the increase of urban area and the management of power and water has become a great challenge for us. We said, though we have acquired control over market due to growth of our economic capacity; corruption, capitalism, the mischievous powers, who are creating imbalance in the market, need more efficiency to be controlled during the 12th plan. He admitted that, the power of inclusion of poor in more incoming group has become slower in spite of rapid economic growth.

It implies that, the rich has become richer by acquiring more wealth. He confessed that, the inflation rate during last three years were high, due to which inclusion of poor in rich group become difficult. If the inflation rate is unbearable, then it creates a hindrance for economic development of the country. It is for information that according to govt. of India the inflation of 5% to 6% and Reserve Bank of India 4% to 5% is bearable. But In the last three years it was much more than that. [4]

Here, however govt. had felt that the development of a few is not the development of the country. Rather development at all levels is highly required. And govt. is framing different policies for the development of public but cannot be successful due to lack of sincerity in their implementation.On the other hand, human resources of Our country are so underdeveloped that, they are unable to know different plans and

---

3. Odia Daily News Paper "The Samaj" dt. 24. 05.2011.
4. Odia Daily News Paper "The Samaj" dt. 24. 05. 2011.

programs of govt. whose opportunities are taken by the corrupt authorities. Such type of culture is found everywhere all over the world, more or less, which was not found before. The position of our country is very much precarious. An assessment made by International Organization regarding corruption where the rank of our country was 82 which is under the rank of Pakistan, whereas our leaders are crying for eradication of corruption with the help of anti-corruption squad. But it is in vain, rather increasing in double or triple form. It means that there is lack of sincerity in such work. It is a cross thinking out for which different kinds of evil things are sprouting from it.

Government is a unitary body, which works for the development of the country as well as for its citizen. But, if different branches will act without keeping coherence among them, the machinery will collapse. There will be corruption, internal and external aggression and state will lose its sovereignty. Our social and political environment will be changed by creating deadlock in the society. Therefore, all the employees, engaged for smooth functioning of the government should follow the rules and regulations framed by it. But idea of becoming millionaire overnight has made the process futile. This is due to cross thinking, i.e., doing and saying are opposite to each other. If they would be honest and virtuous in discharging their duties, no such problems would arise. They should not forget that, they are nurtured and cultured by these people and they are among them, not out of them. If they will act against humanity, what will the future descendents learn? Shall they forgive? The answer is best known to them. Social ecologists view that, environmental problems arise when human beings think themselves out of this environment, otherwise not. Think-tanks should incorporate weak along with strong amidst environment for the beauty of the environment and ecology as a whole.

From this it reveals that, all sorts of mischief are done by human beings in the societies, countries and as a whole in nature. Different rules and regulations are framed by the human beings and violated when and where necessary.

In spite of different moral scriptures and laws, "might is right" policy is followed everywhere. We are all well aware about the environmental degradation, the results that outcome from it and know who are responsible but still not ready to leave our interest. The environmental philosophers of different countries have expressed their deep concern and pointed out that anthropocentricity is the root cause of all sorts of environmental hazards. There are different schools of environmental philosophers or ecologists who coin their theories in different names and propagate how anthropocentricity brings harm to the environment.

Their theories are such as; Deep Ecology, Social Ecology and Eco-feminism which are called Radical Ecology. In traditional ethics, moral standing was granted to human beings only by arguing that, only they have the capacity of judging right or wrong, moral or immoral, which other animal's lack of. But when environmental problems warranted, a new ideological perspective was termed after its biological counterpart i.e., 'Ecology'. In early period, we were not extending moral standing to resolve the environmental crisis. But when the environmental problems became more and more grave, a broader philosophical perspective was required by bringing fundamental change in both our attitude to and understanding of reality. Radical ecologies demand fundamental changes in society and institutions to confront the environmental crisis by changing the way we live and function both as a individual and as a member of society.

DEEP ECOLOGY:

Deep ecology, which comes under radical ecology rejects anthropocentrism and takes a 'total-field' perspective. They give equal status to both living and non-living things. Their view is totally different from swallow ecologists, which is anthropocentric and deals with pollution, resource depletion etc. This swallow deep spilt was first out lined by Norwegian philosopher Arne Naess and advocated for the development of a new eco-philosophy of modern industrial society (Naess, 1973).[5] He and George sessions have compiled eight principles which are basic to deep ecology.

---

5. http://center for deep ecology.org

- The well being and flourishing of human and non-human life on earth have intrinsic values. These values are independent of the usefulness of non-human world for human purposes.
- Richer and diversity of life forms contribute to the realization of these values and have values in themselves.
- Human beings have no right to reduce the richness and diversity, except to satisfy vital needs.
- The flourishing of non-human life requires smaller population.
- The interference of human over non-human world is gradually increasing and the situation is rapidly worsening.
- Present policy regarding environment should be changed which may affect economic, technological and ideological structures.
- The ideological change must be based on the appreciation of life quality or inherent value rather than to the standard of living.
- Those who subscribe to the above points have an obligation directly or indirectly to try to implement the necessary changes (Naess, 1986)

Deep ecologists do not offer one unified ultimate perspective, but possess various philosophical and religious allegiances.

Naess's own ecosophy involves one fundamental ethical norm, i.e., "Self Realization". It means we have to give up the narrow egoistic conception of self in favor of wider and more comprehensive self. By this, he recognizes human being as part of nature, there by identifies human beings with other life forms of nature. This view is also adopted by Australian philosopher Warwick Fox in his eco-philosophy, "Transpersonal Ecology". His philosophy does not hold moral obligation concerning environment, but views it about the realization of an ecological consciousness. Fox told that, if appropriate consciousness will be created, one will automatically protect the environment and allow it to flourish.

SOCIAL ECOLOGY:

It is also a kind of radical ecology, which holds that in order to resolve the crisis, a radical overhaul of this ideology is necessary. The profound social ecologist, Murry Bookchin said that environmental problems are directly related to social problems. In particular, he claims that the hierarchies of power, prevalent within modern societies have fostered a hierarchical relationship between human and natural world. *(Bookchin, 1982)*[6] It is the ideology of free market which has facilitated such hierarchies reducing both human beings and the natural world to commodities. He argues that, the liberation of both humans and nature are dependent on each other. This view is different from Marxist thought, where it is said that man's freedom is dependent on complete domination of natural world through technology. Bookchin's argument is that, humans must recognize that they are part of nature, nor distinct nor separate from it. He holds that no species is more important than another, rather the relation is mutualistic or interrelated. This interdependence and lack of hierarchy in nature provides a scope for non-hierarchical human society. After all, Bookchin does not think that, we should condemn all the humanity for causing ecological crisis; rather it is the relationship within societies that are to blame. But in practice, there is hierarchy in nature. The weak species and weak individuals are killed, eaten and out-competed in an eco-system. This is natural and fits with ecology's characterization of nature as interconnected.

ECO-FEMINISM:

Eco-feminism refers to link between social domination and domination of natural world. It calls for a radical overhaul of prevailing philosophical perspective and ideology of western society. Val Plum wood condemns rationalism that is inherent in traditional ethics, which is responsible for oppression of both women and nature. The fundamental problem is that, it creates dualism. Plumwood says that, traditional ethics promotes reason as capable of providing a stable foundation of moral argument because of its impartiality and universalisability. Emotion lacks these characteristics.

---

6. http://www.social-ecology.org

Plumwood says that, this dualism between reason and emotion grounds other dualisms in rationalist thought; in particular mind/body, human/nature and man/woman. In each case, the former is held to be superior to their later (*Plumwood, 1997*)[7]. So, inferiority of woman and nature has a common source from rationalism.

Many thinkers would argue that, rationalist thought is not harmful, but instead the best hope for securing proper concern for the environment and for women. Karen J. Warren has argued that rationalist thought, as framed by Plumwood is not problematic in itself, but when the question of subordination arises, it becomes problematic. According to Warren, just such logic of domination has been prevalent within western society. Men has been identified with the realm of 'mental' and 'human', while women with 'physical' and 'natural'. When the later was claimed morally inferior, men become justified in subordinating women and nature. Then feminists and environmentalists tried to abolish this oppressive conceptual framework. Other feminists hold that woman is closed to the nature due to their child bearing capacity. Some eco- feminists advocate a spiritualistic approach in which nature and the land are given a sacred value harking back to ancient religions in which the earth is considered female *(Mies &Shiva, 1993)*[8] However, eco-feminists may make the same point as the deep ecologists; to resolve the environmental problems we face and the system of domination in place, it is the consciousness and philosophical outlook of individuals that must change.

It is a bare truth that, one cannot remove oneself from the environment. When it is recognized that, we have environmental obligations, all areas of ethics are affected; like war theory, domestic distributive justice, global distributive justice, human rights theory and so many theories. It can be cited that, Kyoto Protocol was the first real global attempt to deal with the problem of climate change. Though it was not successful, but ethicists cannot remain silent just blaming others, rather propose alternative and better means of resolving the problems we face.

---

7. http://www.good reads.com
8. http://www.amazon.com

For which environmental ethics must be informed by our scientific understanding of the environment that how eco-system works and changes in environmental crisis.

****

CHAPTER-VI

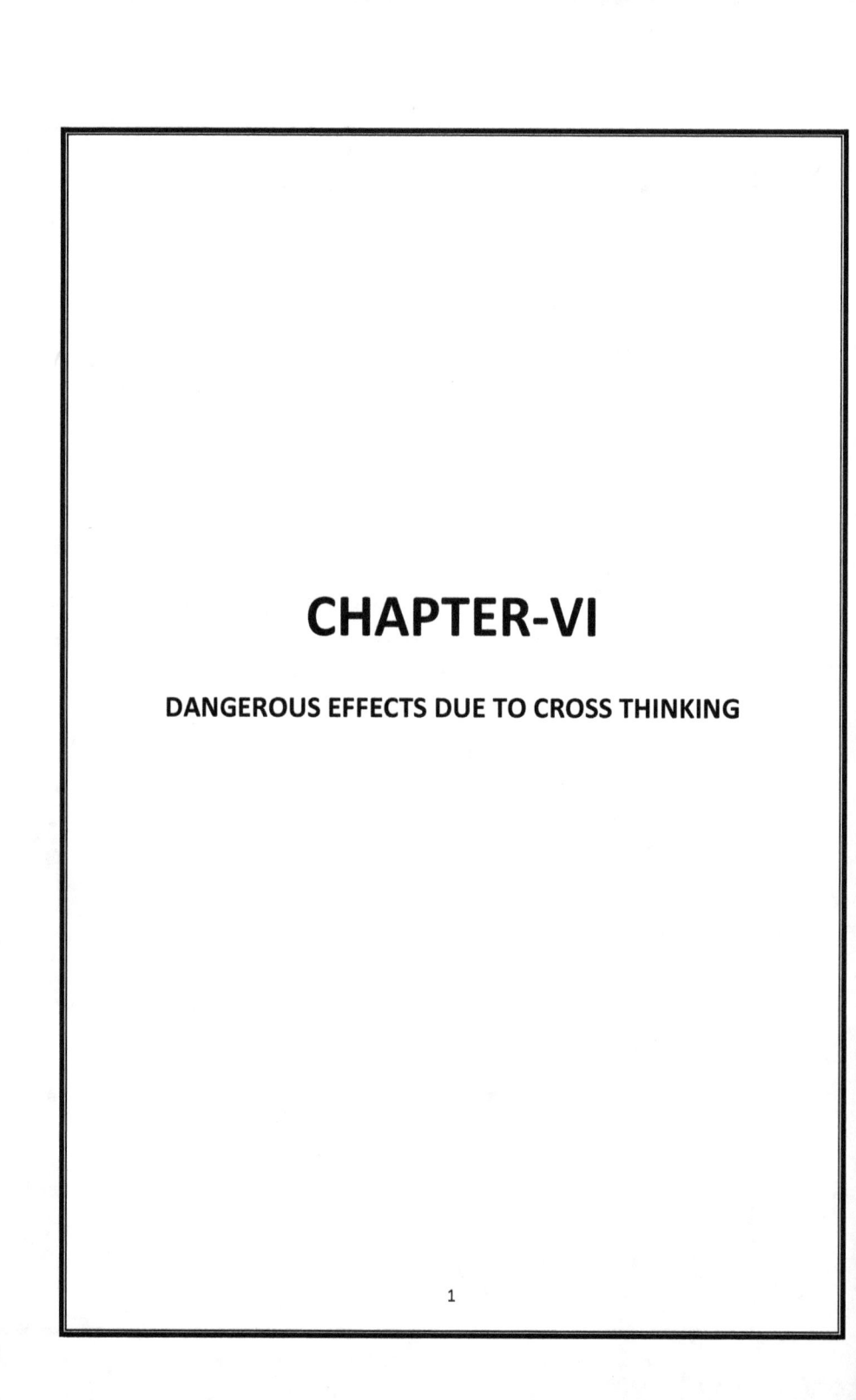

DANGEROUS EFFECTS DUE TO CROSS THINKING

DANGEROUS EFFECTS DUE TO CROSS THINKING

THE EARTH IS ONE; POLLUTION IN ONE PART WILL POLLUTE THE WHOLE:

Dangerous effects arising out of cross thinking are felt to each and every creature of the biotic community. These are all focused in previous chapters during my description. Here, I want to quote the version of Padmashree Dr. Manoj Das, an international writer who had said in a meeting organized in our college campus is like this: A question asked by WHO to the Govt. of India that, "Why people are not happy in spite of all worldly pleasures?" The question was referred to the group of people where doctor, scientist and other intellectuals including Dr. Das, who were the members. They opined that body is not everything and bodily pleasures cannot make us happy, there is something else which is called *Atma*, what our Upanishad had told us long long ago. Atma unites all noble qualities and characters. It is part and parcel of *Viswatma* or *Bramhan* and therefore Indian sages preaches "Vashudhaiba Kutumbakam". All are one. Our great Indian culture says that "Sarve *Vabantu Sukhinah, Sarve Santu Niramaya, Sarve Vadrani Pashyantu, Maa Kaschit Dukhha Vag Vabet*"

But now-a-days, though we are enchanting this in a loud voice in various platforms, but it is just a hollow sound. There is no sincerity in the uttering. Everywhere there is pretention. During the time of worshipping God, we are thinking and praying how to get more for our own. Very few people are thinking for others. We have to remember that, we are living in one world. Whatever demarcation we draw among continent, country, state or private land, those are all man-made, artificial. Those are all part and parcel of the one universe and creation of one creator. We are all enjoying His creation, like other creatures. We are one among others. We have to believe it and accept. The cause of every sorrow is ego-centricity. We the conscious beings should pioneer for the well management of less conscious and unconscious creatures of the universe, keeping in mind that they are the creation of the same creator. This was the thought of our Indian tradition. But when we came in contact with western culture, we began to become more and more materialistic and ran after material pleasures. We oppressed and suppressed nature, so nature is showing its image in turn. When we look into news papers, different types of natural calamities have adorned the pages. No part of the world is left from its cruel raging. It is mostly seen after so called industrial revolution. When human beings felt the degradation or pollution of nature they started to convene meetings for its restoration. Those countries, who pioneer to conduct different meetings to control pollution, they are the most polluters. In 1972, 1974, 1990, 1992, 2002, 2007 and 2010 various meetings are conducted in order to save the mother earth from annihilation for climatic changes.[1] But, result is not up to the point, because of our luxurious life style. The rich nations those who are forming different resolutions cannot obey it due various reasons like political, economic etc. When they are not observing, they cannot compel other small countries which are developing. The essence of the above contention comes like this: The developed countries have cross-thinking, so far as the protection of environment is concerned. They express their thoughts for the protection of environment at the same time; their thoughts when turns into action go directly against the environment, ultimately creating an industrially disturbed society.

THE OUT RAGING OF NATURE IS BEST FELT TO ODISHA:

The ultimate result is that the world is facing various natural calamities. We, being the people of Odisha have felt its raging much better than others. The super cyclone of 1999 and devastating flood of 2004 cannot be forgotten. Again, as Odisha is a state of coastal region, cyclone and flood are its perpetual friends. Due to global warming, the surface of the sea water is becoming hot. So, high waves are coming; storm, cyclones are formed. Depression frequently occurs, which brings heavy rain fall, cyclone etc and cause great loss to our state. All the countries situated on the sea coasts are suffering a lot by the out raging of the nature.

But in spite of all the bad effects caused by global warming due to industrialization and other co-relating activities, Govt. (both central and state) are permitting different national, multinational companies to set up industries in the name

---

1. Odia daily News Paper "The Samaj", dt. 24. 05. 2011.

of economic development. Here, I want to cite POSCO, Vedanta etc. They will use our fertile land, water, jungle, mines and all the natural resources by displacing our own people who have been enjoying these things on ancestral basis. In spite of their strong protest, govt. is trying to take away their rights of property in order to offer those in the hands of those multinational companies. Different thinkers have expressed their opinion regarding loss and gain and suggested not to permit these companies, because our people, our state will lose more, but get less. If jungles will be cut down, our valuable bio-diversity will be lost and the indigenous people living in and depending on will die out of starvation. The environment will be more and more degraded. Now-a-days the temperature of Talcher, Titilagarh is raising up to 50^0 Celsius during summer. If all the MOUs will be materialized, what will be the temperature and what will be the condition of the environment of our state. Time will come when monsoon air cannot rain. The water vapour will be evaporated in the sky and Odisha will be converted into desert. What will be the meaning of economic development? So, it is not for the upliftment of the people of Odisha, but for the gigantic gain of those multinational companies and certain vested interest politicians and bureaucrats. In the name of economic development, the govt. is dragging the innocent and peace loving people of Odisha into the cavity of various pollutions like forcibly dragging of cattle into slaughter house.

NOT RANDOM USE OF NATURAL RESOURCES; BUT MAN-MADE RESOURCES TO BE DEVELOPED FOR ECONOMIC GROWTH:

Here, I would like to suggest that excessive use of non-renewable resources in the name of economic development cannot bring the permanent solution to the economic problems. Steps should be taken by framing different policies and schemes for the development of the man-made resources; they will be capable to earn their livelihood by using their intellect. The countries which are called developed, only due to development of their human resources; in this case we have to follow the developed countries. Government is also sending diplomats and bureaucrats to different countries to know about their techniques of development in different fields, but at the same time we have to follow how honestly they are performing their duties. Our govt. is also preparing various schemes for the wellbeing of our people, but they are becoming unsuccessful due to certain corrupt authorities. That is the dangerous effect arising out of cross-thinking. So, many people are debarred from getting the good results of different programs.

SCIENTIFIC INVENTIONS ARE FOR THE BETTERMENT OF THE SOCIETY, NOT FOR DESTRUCTION:

Now it is seen that, if something is newly made or invented for the betterment of human race, immediately its misuse is thought of. The inventor who has taken so much of pain for it will get hurt. Nuclear energy was invented for peaceful implementation and for the benefit of the human race. But political leaders have used it for preparation of bombs for invasion of countries, i.e.; for the destruction of human and non-human beings. Terrorists are trying to take the privilege of it. Was it the mission of inventor? Certainly not. Likewise, the outcome of sophisticated devices has direct impact on corruption, adulterations etc. These devices help to accelerate immoral activities. In the name of mental refreshment, our people watch different lustful programs of various cultures in TV and other audio visual aids, which are prohibited in our society and try to practise these among their neighboring people and relations which are criminal offence in our society and create law and order situation. It attracts the young mass very much. It is also the human tendency to accept and learn bad things more quickly than good things. Therefore, our young generations are gradually deteriorating from the moral point of view and committing different types of crimes which adorn the pages of news papers daily. The dresses worn by artists and their activities create sensation among the young generation and provoke to commit crimes like rape, murder, sophisticated theft, stealing from ATM etc which are not heard before. These are the sheer outcome of cross thinking.

HUMANLY QUALITIES ARE RARELY FOUND DUE TO CROSS CULTURAL THINKING:

The cross thinking has made our mind so perturbed that, we are all trying to get rid of it. Everywhere, there is unrest in the society. Because, we all are trying to get more with in a very short period at any rate. Truthfulness, dutifulness, loyalty, patriotism are rarely found in the society. Those who are elected to rule the public; they take away the public money. The judges, those who are called as *Dharmavatara,* they are doing unjust work by taking bribe or so. Pollution and corruption have covered the earth. Our governments are saying for prevention of corruption, but the government policies are promoting high rate of corruption. Those who are raising voice against corruption, they are put to harassment. They get enormous public support to fight against corruption but government does not respond to their demand. Therefore, learned and honest people are less interested to take part in politics. If good and honest men will not enter into it, how shall we expect good administration? If administration will not good, country cannot prosper and it will remain as it is or go back or that what is going in West Asia. Our progress is also not satisfactory; it would be much more if there would be no corruption in our country. When the society is disturbed and economically sabotaged, people will be compelled to destroy environment for their livelihood.

THE GLOBE IS IN THE CAVITY OF HORROR:

Another thing which each and everybody feels that, world is on the lap of explosives and nobody knows whenever it will embrace us. We have invented sophisticated weapons for our destruction, i.e., destruction of other country in order to satisfy our ego, arrogance. If we will do, why shall not they? So, all the countries have stored different sophisticated weapons in the name of their own safety. Any kind of mismanagement can destroy the whole or part of it. The fear of terrorism is also there. If they will get these weapons or its technology, then result cannot be thought of. We are creating these things and advising others not to prepare these things. Shall it be possible? Definitely 'No'. Therefore we are happy in one hand but trembling with fear on the other. This is the outcome of the cross thinking. We are saying that global warming is causing various problems like Tsunami, heavy rain fall, drought, devastating flood, super cyclone, earth quake etc. which are occurring in different parts of the world. Global warming is chiefly caused due to burning of mineral oil, coal for running of vehicles, factories and production of power. Here a question comes to my mind. Do governments impose any control regarding production of vehicles? Then arise the question of using vehicles, on the other hand, companies are producing different luxurious models and advertising in the market on competition basis. Though, they certify that they are less pollutant, but their increased numbers have surpassed the old pollutant vehicles. The same is in case of industries. So, we the human beings are digging the graveyard of our own as well as other lives on the earth.

DIFFERENT HEALTH HAZARDS ARE CREATED DUE TO IMPACT OF OTHER CULTURE:

We the human beings always want to lead a luxurious life. Our food habit, reluctant to do physical labour, unfaithful to life partner invite various incurable diseases like diabetes, high pressure, heart problem, cancer, AIDS and other such diseases. The more we become luxurious, the ratio of these diseases are increasing. Though we know that fast foods are not good for health, still we are using it. Our native foods are gradually going away from breakfast. Non-vegetarian foods are mostly liked by young mass, which are the root of many diseases. In India, faithfulness and truthfulness are the canon of binding a couple. But it is now lessened. These are all due to impact of western culture. Therefore, different types of diseases are found in our society. But it is a good sign that, many people are practising yoga therapy of Indian origin to make themselves fit, both inside and outside the country. Our old sages had kept everything for the betterment for their ancestors. But we neglect them saying old, outdated and follow the westerners. When we face a lot of problems and when they recognized our Indian life style, herbal medicines, yoga as rights; then we say that these are ours, our predecessors have prescribed and become proud of it.

BLIND SUPPORT OF FOREIGN CULTURES HAS BROUGHT SO MANY BAD THINGS TO OUR CULTURE:

Here one thing I have marked, that when something is recognized by the western people as good, we accept it without thinking the pros and cons of it. Therefore, they are always bargaining us. Our past Indian culture, tradition, vogue, folk stories; all are much more valuable than others for us. The stories of the *Ramayana* and *Mahabharat*, the sermons of *Sri Krishna* to *Arjuna* in *Bhagabat Gita*, the advices of *Bhagabat* are incomparable to any other epics of the world. When people of other languages knew them, they could not stay without praising it as well as our Indian culture. Our ancestors have created these as they are suited to our environment. But when we followed the foreign people; their culture, tradition, food habit, dress code etc, our culture became polluted but our young generations become proud of adopting foreign cultures.

Especially, three things which are very much antagonistic to our culture, i.e.; food habit, dress code and unfaithful to life partner. Now our young people are eagerly eating different fast foods like *Pizza, Berger* and different Chinese items which include both veg. and non-veg. items. In non-veg. category the germs in the animal flesh enter into our body and create various diseases which were present among those animals. Likewise; dresses are worn to cover our body. But now our young girls are wearing different types of so short and slim dresses that they create excitement or lust among the opposites and society faces so many problems. The present society is over burdened with such type of problems. In ancient time, the ladies of different ages were honoured and various festivals are conjoined to grace those occasions like *Raja, Kumar Purnima, Dasahara, are* observed in order to recognize the superiority of female. Another thing, which is pervading like epidemic, that is unfaithful to co-partner. Not a single day is left where we do not find woman harassment in the news paper. Different types of sexually transmitted diseases like AIDS are pervading in the society where death is inevitable. In some cases, innocent partners are suffered. This is how cross thinking has endangered our lives. Our ancestor Sri Achyutananda Das had predicted the sufferings of human life before five hundred years in his "Malika Bachan". He had written that:-

"Baishi Pahache Kheliba Mina"

This means, fishes will play in the twenty two steps of Lord Jagannath temple at Puri. Now we see that, due to global warming sea level is increasing. Therefore, the coastal belts will be drowned, twenty two steps of Lord Jagannath temple will be drowned and fishes will play there. Such type of prediction was made by our ancestor about the result of global warming.

It is proved that, there is everything in our culture, heritage and tradition. What is the necessity of following others when ours have such a glorious heritage? Following others blindly, we are inviting dangers to us. Directly this type of tendency has a bearing on the environment propelling the man to be reckless towards environment.

THE UNWANTED INTEREST FOR POSCO MAY BRING DANGEROUS RESULTS IN VARIOUS SECTORS:

The people of odisha are peace loving and they like to live peacefully in spite of acute poverty. Cultivation is the main source of income of majority of people. But when different industries will be set up, the temperature will rise, river water and underground water will go down; the food, water, air everything will be contaminated, the local people, domestic animals, wild animals will suffer a lot and public agitation will grow up against industrialization. The situation in different districts like Jajpur, Keonjhar, Jharsuguda, Angul and some others are very much precarious. During the time of my project work POSCO agitation is continuing. The people of that area are fighting for their livelihood even if they are involving their children to save their lands from forcible action taken by govt. Though POSCO is agreed to give appointment to one member from a family who are losing land, native people are not accepting it. They say that POSCO will not give appointment on ancestral basis, but they are enjoying the land on ancestral, basis. Again, as it is a multinational company it will look its own profit. He will never bother about the interest of our people, rather use different modern technologies and terminate workers for more profit. As local people are non-technical men, they will be appointed at low cost salary. Company will recruit technical persons from outside, who are experts, the executive posts will remain in the hands of company, where the voice of the local people will be dominated. So, company, by using our raw material, land, water, port, SEZ facility will become more and more rich by leaving various pollutions, health hazards and environmental degradation for our people. During the ministry of Mr. Biju Pattanayak, he has requested the POSCO to set up steel industry in Odisha as his mission was to establish one thousand industries in one thousand days. But POSCO did not pay any heed to it. But now POSCO is showing much interest, because at that time the cost of iron was very low in world market, but now it has risen too much. So the profit will be sky rocketed. Therefore, POSCO is showing so much interest. However, the people of that area can imagine their future, for which they are opposing so staunchly. If at all, the modern man cries for environmental issues, what is the necessity of destroying so many cultivable lands for the sake of multinational companies? Rather, factories should be established far away from the villages and native places.

These are few citations of cross thinking. There are so many examples of cross thinking which tortures our lives every now and then. No individual effort can eradicate it, combined effort or societal effort is required to keep our old traditions, styles intact. Otherwise, they will be mingled in the super cyclone of globalization. Nobody is happy now. Everywhere there is cry for peace. In spite of so much physical gain than before, we are not fully fledged. We are eating tasteless, writing charmless and thinking heartless. The meaning is that we are doing action having no passion. So we are leading a heartless life, where everything is done mechanically.

****